# Synthesis Lectures on Welding Engineering

AF803699

**Series Editors**

Menachem Kimchi, State Library of Ohio, Columbus, USA

David H. Phillips, Columbus, USA

This series publishes short books on fundamentals, principles, and applications on a variety of Welding Engineering topics including spot welding, solid-state welding, a guide to navigating welding codes, computational modelling of welds and welding processes, laser welding, arc welding automation and control, and micro-joining. Initial topics have been chosen because they are either not well represented by the current market of published material, and/or similar publications are outdated and do not represent the latest welding advancements, knowledge, and technology.

Darren Barborak

# Arc Welding Qualification Standards

## Fundamentals and Application

Springer

Darren Barborak
The Ohio State University
Columbus, OH, USA

ISSN 3005-0340 ISSN 3005-0359 (electronic)
Synthesis Lectures on Welding Engineering
ISBN 978-3-031-64648-5 ISBN 978-3-031-64646-1 (eBook)
https://doi.org/10.1007/978-3-031-64646-1

© The Editor(s) (if applicable) and The Author(s), under exclusive license to Springer Nature Switzerland AG 2025

This work is subject to copyright. All rights are solely and exclusively licensed by the Publisher, whether the whole or part of the material is concerned, specifically the rights of translation, reprinting, reuse of illustrations, recitation, broadcasting, reproduction on microfilms or in any other physical way, and transmission or information storage and retrieval, electronic adaptation, computer software, or by similar or dissimilar methodology now known or hereafter developed.
The use of general descriptive names, registered names, trademarks, service marks, etc. in this publication does not imply, even in the absence of a specific statement, that such names are exempt from the relevant protective laws and regulations and therefore free for general use.
The publisher, the authors and the editors are safe to assume that the advice and information in this book are believed to be true and accurate at the date of publication. Neither the publisher nor the authors or the editors give a warranty, expressed or implied, with respect to the material contained herein or for any errors or omissions that may have been made. The publisher remains neutral with regard to jurisdictional claims in published maps and institutional affiliations.

This Springer imprint is published by the registered company Springer Nature Switzerland AG
The registered company address is: Gewerbestrasse 11, 6330 Cham, Switzerland

If disposing of this product, please recycle the paper.

# Preface

This textbook introduces the reader to the development and qualification of arc welding procedures and personnel to industry codes and standards. The mechanics of using welding standards, how to address their requirements, and their relationship with other standards are explained. The reader will gain a working knowledge of common welding standards including a review of welding processes variables, the inspection and testing of welds, and their acceptance criteria.

The reader will develop a basic understanding of:

- Common arc welding standards.
- Welding-related documentation.
- The welding procedure development and qualification process.
- The welder/operator performance qualification process.
- Essential, non-essential, and supplementary essential variables for arc welding processes.
- The requirements for the inspection and testing of weld qualification coupons.
- Purpose, intent, and compliance of a Welding Procedure Specifications (WPS).
- Purpose, intent, and compliance of a Procedure Qualification Records (PQR).
- Purpose, intent, and compliance of a Welder Performance Qualification Record (WPQR).

This textbook was written for use in an undergraduate course in Welding Engineering that I have taught at The Ohio State University for the past 10 years. Much of what is included in this book comes from my class lectures. Although the book is aimed at Welding Engineering students, it should also serve as a useful guide to other engineers, technicians, and specialists who are working in the field of welding and are seeking how to apply relevant codes and standards to qualify welding procedures and personnel. While the book focused primarily on the common arc welding processes using AWS B2.1 and

ASME BPVC Section IX, the principles discussed will apply to most welding processes in general and most welding qualification standards.

It is recommended that the reader has access to AWS B2.1 *Specification for Welding Procedure and Performance Qualification* 2021 edition and/or ASME Boiler and *Pressure Vessel Code ASME IX Welding, Brazing, and Fusing Qualifications* 2023 edition as this book frequently references paragraphs, clauses, and figures from these standards. The manuscript was written using U.S. Customary units as the primary measurement units followed by Si units in brackets, i.e., "[ ]". ASME BPVC IX Mandatory Appendix G was utilized for guidance in converting between measurement units.

Columbus, OH, USA Darren Barborak

# Acknowledgements

The author would like to acknowledge the series editors, Profs. David Phillips and Menachem Kimchi, for their guidance in preparing this manuscript, the helpful peer reviews from friends and professional colleagues, as well as the students from my Welding Engineering 4602 classes at The Ohio State University. Finally, I would like to acknowledge the patience of my parents, who fostered my curiosity with all things mechanical and electrical which turned into a career in engineering, and to my loving wife Antoinette, whose unwavering support allowed me to complete this manuscript.

# Contents

# Abbreviations

| | |
|---|---|
| AI | Authorized Inspector |
| ANS | American National Standard |
| ANSI | American National Standards Institute |
| API | American Petroleum Institute |
| ASME | American Society of Mechanical Engineers |
| ASTM | American Society for Testing and Materials |
| AWS | American Welding Society |
| BPVC | Boiler and Pressure Vessel Code |
| CEN | European Committee for Standardization |
| CMTR | Certified Material Test Report |
| CRO | Corrosion Resistant Overlay |
| CWB | Canadian Welding Bureau |
| DE | Destructive Examination |
| FCAW | Flux Cored Arc Welding |
| GMAW | Gas Metal Arc Welding |
| GTAW | Gas Tungsten Arc Welding |
| ISO | International Organization for Standardization |
| MT | Magnetic |
| MTC | Mill Test Certificate |
| MTR | Material Test Report |
| NBIC | National Board Inspection Code |
| NDE | Non-Destructive Examination |
| NDT | Non-Destructive Testing |
| NIST | National Institute of Standards and Technology |
| NSB | National Standards Body |
| PAW | Plasma Arc Welding |
| PQR | Procedure Qualification Record |
| PQTR | Performance Qualification Test Record |

| | |
|---|---|
| PWHT | Post-Weld Heat Treatment |
| PWPS | Preliminary Welding Procedure Specification |
| SAW | Submerged Arc Welding |
| SDO | Standard Development Organization |
| SMAW | Shielded Metal Arc Welding |
| SSO | Standards Setting Organization |
| SWPS | Standard Welding Procedure Specification |
| TAC | Technical Activities Committee |
| WOPQ | Welder Operator Performance Qualification |
| WPQ | Welder Performance Qualification |
| WPS | Welding Procedure Specification |

# Introduction

# 1

In 1904, a fire broke out at the John E. Hurst & Company Building in Baltimore Maryland. After consuming the entire structure, it leaped from building to building until it engulfed 2,500 buildings in an 80-block area of the city over 30 h. Fire fighters from New York, Philadelphia, and Washington, DC responded to help battle the blaze, but unfortunately, they had to watch helplessly because their hoses could not connect because they did not fit the hydrants in Baltimore (Fig. 1.1).

Up until that time, each municipality had its own unique set of standards for firefighting equipment. It was evident that a new national standard had to be developed to prevent a similar occurrence in the future. As a result, the National Fire Protection Association (NFPA) formed a committee and research was conducted of over 600 fire hose couplings from around the country. One year later, a standard was created to ensure uniform fire safety equipment and the safety of Americans nationwide. This standard specified that each fire hydrant have one large diameter port 4.5 inches in diameter with 4 threads per inch (meant for supplying water to a pumper truck or other high-capacity distribution device), and two medium-diameter ports, each 2.5 inches with 7.5 threads per inch, meant for supplying individual hoses directly. This standard would eventually become an official standard, NFPA 1963.

Because this standard is considered a "voluntary" standard, its adoption is not mandatory. To this day, there remain significant municipalities in North America that use fire hose and hydrant threads and other couplings that are incompatible with those used by neighboring fire departments. These incompatibilities have led to well-documented loss of life and structures such as the Oakland Firestorm of 1991 (also known as the Tunnel Fire, Oakland Hills firestorm, or East Bay Hills fire). The 1,520-acre fire caused $1.5 billion in damage, destroying 2,699 single-family dwellings, and ultimately 25 people perished. To

© The Author(s), under exclusive license to Springer Nature Switzerland AG 2025

D. Barborak, *Arc Welding Qualification Standards*, Synthesis Lectures on Welding Engineering, https://doi.org/10.1007/978-3-031-64646-1_1

**Fig. 1.1** Aftermath of the 1904 Baltimore Fire

this day, many fire hydrants in San Francisco are incompatible with the equipment used by many nearby fire departments that would respond in mutual aid situations.[1]

## 1.1 History of Standards

Standards have existed since the beginning of recorded history. Some were created by royal decree. Some standards were an outgrowth of man's desire to harmonize his activities with important changes in the environment. Other standards were created in response to the needs of an increasingly complex society.

The process of standardization has set the stage for immense gains in collaboration, productivity, and innovation. Standards allow us to find collective harmony within a society that grows increasingly complex.

Standardization of Time

One of the earliest examples of standardization of time is the creation of a calendar. Ancient civilizations relied upon the apparent motion of the sun, moon, and stars through the sky to determine the appropriate time to plant and harvest crops, to celebrate holidays, and to record important events.

Over 20,000 years ago, our Ice Age ancestors in Europe made the first rudimentary attempts to keep track of days by scratching lines in caves and gouging holes in sticks and bones. Later, as civilizations developed agriculture and began to farm their lands, they needed more precise ways to predict seasonal changes.

The Sumerians in the Tigris/Euphrates valley devised a calendar very similar to the one we use today. 5,000 years ago, the Sumerian farmer used a calendar that divided the year into 30-day months. Each day was divided into 12 h and each hour into 30 min.

The Egyptians were the first to develop the 365-day calendar and can be credited with logging 4236 BC as the first year in recorded history. They based the year's measurement on the rising of the "Dog Star" or Sirius every 365 days. This was an important event as it coincided with the annual inundation of the Nile, a yearly occurrence that enriched the soil used to plant the kingdom's crops.

The idea of using atomic transitions (a change of an electron from one energy level to another within an atom) to measure time was suggested by Lord Kelvin in 1879. The first implementation of an atomic clock was in 1949, but the technology continues to evolve today. Presently, the accuracy for an atomic clock is expected to neither gain nor lose a second in about 138 million years.

Standardization of Length

About 8,000 years ago, the people of the first civilizations along the Tigris, Euphrates, and Nile river basins used their own body parts such as their forearm, hands, and fingers as standards to build huts, looms, plows, hoes, and sickles.

About 5,000 years ago, the Egyptians produced the first standard of linear measurement known as the Royal Cubit. This standard was made from black granite and measured 523–525 mm (20.6–20.64 inches) in length. The length of the sides of the Great Pyramid varies by no more than 0.05 percent because of the use of this standard. 2000 years later, the Greeks developed their own standard based on two-thirds the size of the Royal Cubit with 16 equal divisions.

Around 900 years ago, King Henry I of England instituted the ell based on the length of his arm. This later led to the Iron Ulna known as the British standard yard. 200 years ago, the Meter of the Archives was developed out of a platinum bar with a rectangular cross section and polished parallel ends. The meter was defined as the distance between the polished end faces at a specified temperature and it was the international standard for most of the 19th century.

In 1898, C.E. Johannson filed a patent for a set of gauge block standards which could be combined into many different lengths. Calibrated against a master set of blocks and ultimately against the international standard of length, these gauge blocks became each shop's link to a universal standard.

The meter was ultimately defined by international agreement in 1983 as the length of the path traveled by light in a vacuum in 1/299,792,458 of a second. Length is now no longer an independent standard but rather is derived from the extremely accurate standard of time and a universally agreed upon defined value for the speed of light.

Industrial Revolution

Indeed, it was broad standardization that paved the way for the Industrial Revolution. Interchangeable parts dramatically reduced costs, allowing for easy assembly of new goods, cheap repairs, and most of all, they reduced the time and skill required for workers. Or consider how those manufactured products are then shipped—likely by train. Prior to the standardization of the railroad gauge, cargo traveling between regions would have to be unloaded and moved to new trains because the distance between rails no longer matched the train's wheels.

With the advent of the Industrial Revolution in the 19th century, the increased demand to transport goods from place to place led to advanced modes of transportation. The invention of the Railroad was a fast, economical, and effective means of sending products cross-country. This feat was made possible by the standardization of the railroad gauge, which established the uniform distance between two rails on a track. Imagine the chaos and wasted time if a train starting out in New York had to be unloaded in St. Louis because the railroad tracks did not line up with the train's wheels. Early train travel in America was hampered by this phenomenon.

During the Civil War the U.S. government recognized the military and economic advantages to having a standardized rail track gauge. The government worked with the railroads to promote use of the most common railroad gauge in the U.S. at the time which measured 4 feet, 8 1/2 inches, a track size that originated in England. This gauge was mandated for use in the Transcontinental Railroad in 1864 and by 1886 had become the U.S. standard.

## 1.2 What is a Standard?

According to the US Government Office of Management and Budget,[2] the term "standard" includes all of the following:

1. Common and repeated use of rules, conditions, guidelines or characteristics for products or related processes and production methods, and related management systems practices
2. The definition of terms; classification of components; delineation of procedures; specification of dimensions, materials, performance, designs, or operations; measurement of quality and quantity in describing materials, processes, products, systems, services, or practices; test methods and sampling procedures; formats for information and communication exchange; or descriptions of fit and measurements of size or strength
3. Terminology, symbols, packaging, marking or labeling requirements as they apply to a product, process, or production method.

Standards impact everyday life by facilitating trade, commerce, and innovation; reducing costs in the public and private sectors; supporting interconnectivity of products and systems; keeping people safe by minimizing injuries and protecting key environmental resources; and facilitating the advancement of scientific discovery.

Standards can be broadly categorized by their use, application, or intent as follows:

- Product standards: banking cards, washing machines
- Safety standards: lifejackets, eyewear, boiler pressure vessel code, national electrical code
- Performance standards: food and toy safety
- Management systems standards: quality, environmental and energy management
- Personnel standards: food handlers, crane operators, welders
- Information and Communication standards: underpin almost everything
- Construction standards—Standards which provide requirements, information, and guidance for the fabrication of an item prior to it being used in service.
  - AWS D1.1, Structural Welding—Steel
  - ASME BPVC VIII, Pressure Vessels.
- Repair standards—Standards which provide requirements, information, and guidance for the inspection, repair, or alteration of an item after it has been placed in service.
  - ASME PCC-2, Repair of Pressure Equipment and Piping
  - NBIC NB-23 Part 3, Repairs and Alterations
  - API 510, Pressure Vessel Inspection Code: Maintenance, Inspection, Rating, Repair, and Alteration.
- Service and Support standards—Standards invoked or specified by a Construction or Repair Standard, or for use with a specific application.
  - ASME BPVC IX, Qualification Standard for Welding, Brazing, and Fusing
  - AWS B2.1, Specification for Welding Procedure and Performance Qualification.

A technical standard establishes uniform engineering or technical criteria, method, process, practice, or procedure to assure that minimum requirements of quality and safety are met in the manufacturing or construction of materials, products, and structures. The following describes the different types of technical standards.

### 1.2.1 Standard Specifications

A Standard Specification clearly and accurately describes the essential technical requirements for a material, component, product, system, or service. Specifications are used heavily by purchasing departments for controlling the quality of incoming materials. A specification is intended to be mandatory when so required.

A Requirement Specification is a documented requirement, or set of documented requirements, to be satisfied by a given material, design, product, service, etc. It is a common early part of engineering design and product development processes, in many fields.

A Functional Specification is a kind of requirement specification and may show functional block diagrams.

A Design or Product Specification describes the features of the solutions for the Requirement Specification, referring to either a designed solution or final produced solution. It is often used to guide fabrication/production. Sometimes the term specification is here used in connection with a data sheet (or spec sheet), which may be confusing. A data sheet describes the technical characteristics of an item or product, often published by a manufacturer to help people choose or use the products. A data sheet is not a technical specification in the sense of informing how to produce.

An Operational or Performance Specification, specifies the conditions of a system or object after years of operation, including the effects of wear and maintenance (configuration changes).

Examples of welding related specifications include:

- AWS A5.02 Specification for Filler Metal Standard Sizes, Packaging, and Physical Attributes
- ASME SA-516 Specification for Pressure Vessel Plates, Carbon Steel, for Moderate and Lower Temperature Service
- ISO 14175 Welding Consumables—Gases and Gas Mixtures for Fusion Welding and Allied Processes.

### 1.2.2 Standard Guides

A Standard Guide provides information options to the user as to the best practical methods to perform a given task but do not require a specific course of action. A guide is not intended to be mandatory, unless they are referenced in codes or contractual agreements.

Examples of welding related Guides include:

- AWS B1.11 Guide for the Visual Inspection of Welds
- ISO 18491 Welding and allied processes—Guidelines for measurement of welding energies.

### 1.2.3 Standard Practices and Procedures

A Standard Practice or Procedure describes general industry practice by providing a set of instructions for a particular process, material, technique, or method. They are not intended to be mandatory, unless they are referenced in codes or contractual agreements.

Examples of welding-related Recommended Practices include:

- AWS A4.2 Standard Procedures for Calibrating Magnetic Instruments to Measure the Delta Ferrite Content of Austenitic and Duplex Ferritic-Austenitic Stainless Steel Weld Metal
- AWS C4.2 Recommended Practices for Safe Oxyfuel Gas Cutting Torch Operation
- ISO 3690 Welding and allied processes—Determination of Hydrogen Content in Arc Weld Metal.

### 1.2.4 Standard Methods

A Standard Method consists of a set of requirements relating to the way a particular kind of test, sampling, analysis, observation, or measurement is conducted to determine properties, composition, or performance of some item. Methods are not intended to be mandatory unless they are referenced in codes or contractual agreements.

Examples of welding-related methods include:

- AWS A4.0 Standard Methods for Mechanical Testing of Welds
- ASTM E94 Standard Test Method for Vickers Hardness of Metallic Materials
- ASTM E165 Standard Test Method for Liquid Penetrant Examination.

### 1.2.5 Standard Terms and Definitions

A Standard for Terms and Definitions formally establishes terminology, units, accuracy Standard units, in physics and applied mathematics, are commonly accepted measurements of physical quantities.

Examples of welding-related Terms and Definitions include:

- AWS A2.4 Standard Symbols for Welding, Brazing, and Nondestructive Examination
- AWS A3.0 Standard Welding Terms and Definitions
- ISO 857-1 Welding and allied processes—Vocabulary—Part 1: Metal welding processes
- ISO 2553 Welding and allied processes—Symbolic representation on drawings—Welded joints.

## 1.3 Standards Development and Maintenance

The process of developing a standard is typically facilitated by a Standards Development Organization (SDO) also known as a Standards Setting Organization (SSO). The SDO has the primary responsibility of the standards development lifecycle including developing, coordinating, promulgating, revising, amending, reissuing, interpreting, or otherwise producing standards.

SDOs can be generally classified as:

- Government (Local, State, Federal)
- Industry (Private Companies and Organizations, Consortia, Non-Profits)
- Geography (Local, National, Regional, and Global).

Each SDO applies its own processes, rules, terminology, and format to the standards development process. Typically, each SDO is comprised of Boards, Groups or Committees, and staff who establish and maintain the policies, procedures and guidelines that help ensure the integrity of the standards development process, and the standards that are generated as an outcome of this process.

The development of a new standard is typically triggered by a formal request, submitted to an SDO by an individual or entity, such as an industry society or consortia, for consideration. Once the SDO approves the request to develop a new standard, a collaborative team of volunteers is assembled. Depending on the SDO, this collaborative team can be called a Subcommittee, Working Group, or Task Group and is comprised of individuals and/or entities (companies, organizations, non-profits, government agencies), all having an expertise in the particular subject matter of the standard. Collectively, these volunteer participants carry a specific interest in a specific area of development as producers, sellers, buyers, users and/or regulators of a particular material, product, process, or service. Participants may contribute at varying levels to the standards development process, depending on the rules and criteria of the SDO. These rules help ensure that no one single interest dominates the standards development process.

When a standards development group is formed, officers may either be elected by the group members or appointed by the SDO. Officers oversee the standards development project and its adherence to the SDO process and rules. The standards development group may establish their own individual, organizational, communications and meeting structures, and govern work process, activities, consensus building, decision making, balloting and even financial reporting in accordance with SDO rules.

To build consensus through democratic means, participants engage in meetings, draft and review position pieces, create and review presentations, examine data and engage in active discussion and debate to resolve outstanding issues. Eventually, a draft standard is compiled which may undergo multiple revisions. During this process, the draft standard may be published for public comment. Once a draft standard has been finalized by the standards development group, it may be submitted for broader consideration and balloting within the SDO such as its Standards Committee, Main Committee, or even Board of Directors. After submission, review and acceptance, the approved standard is published and made available for distribution and purchasing within several outlets, including through the SDO itself.

Standards are considered living documents, which may be iteratively modified, corrected, adjusted and/or updated based on industry/market conditions and other factors. At any given point in time, a standard may be referred to as having multiple "status" classifications which may include:

- Approved Project—An initial standards project request is approved, and in stages of group formation.
- Active Project—An active standards development project in-progress.
- Withdrawn Project—A cancelled standards development project.
- Approved Standard—The standard is approved and published for public use.
- Withdrawn or Annulled Standard—The standard is no longer relevant or active.
- Reaffirmed Standard—The standard has been reviewed and determined to be current with no need for immediate revision.
- Revised Standard—The standard has been updated and republished. These updates may include editorial changes, revisions of the rules, new rules as dictated by technological development, and inclusion of new Code Cases and interpretations.
- Superseded Standard—The standard has been replaced by a new standard in its entirety.

The review cycle timeline for the different SDOs varies by organization. For example, AWS standards and ISO standards are reviewed every 5 years while the ASME boiler and pressure vessel standards are reviewed every 2 years.

In between the review cycle, additional supporting documentation may be released to communicate changes or corrections that will eventually be incorporated into the revised standard. These supporting documents include:

- Amendment—A document that adds to, removes from, or alters material in a portion of an existing standard and may make editorial or technical corrections to that standard.
- Corrigendum—A document that only corrects editorial errors, technical errors, or ambiguities in an existing standard. A corrigendum does not introduce new material.
- Erratum—A document that contains only grammatical corrections to, or corrections of errors introduced during the publishing process of, an existing standard. An erratum is based on the comparison of the final balloted version of the standard as compared to the published version.

In keeping with the standards development life cycle, the group that developed the standard may also go through periodic stages of activity or dormancy. Depending on where a standard is in its life cycle a standard may be accompanied by supplemental documents including Errata (which address errors in the standard), Amendments (which modify sections of the standard), Corrigenda (which only correct errors or ambiguities in a standard), handbooks, reports, tutorials, and other related materials. These supplemental documents help users better understand and apply the standard.

To educate the user in what changes were made from the previous version, a standard may include a Synopsis or Summary of Changes describing the changes. In addition, many SDOs require a vertical line in the margin and/or underlined text in clauses, tables, or figures to indicate an editorial or technical change.

### 1.3.1 Interpretations

Interpretations are replies to inquiries concerning the interpretation of technical aspects of the Code made by its Standard Development Organization (SDO) staff or committee members. A response to an inquiry may be formal or informal (Fig. 1.2).

- Formal written responses to written inquiries which are transmitted to the inquirer on organizational letterhead. Formal Inquiries are often compiled and published by the SDO.
- Informal responses to inquiries may be offered by staff and volunteers. Such individual responses should be accompanied by a statement making it clear that they are the

**Standard Designation:** BPV Section IX

**Edition/Addenda:** 2021

**Para./Fig./Table No:** QW-408

**Subject Description:** BPV IX-2021: QW-408, Backing gas purity

**Date Issued:** 01/17/2024

**Record Number:** 22-1720

**Interpretation Number:** BPV IX-24-03

**Question(s) and Reply(ies):**

Question: For a single backing gas or combination of backing gases, is a change in the purity of the backing gas an essential variable?

Reply: No.

**Fig. 1.2** Example interpretation for ASME Section IX

opinion of the individual, not interpretations. These responses may be either verbal or written. If written, the responses shall not be on organizational letterhead.

### 1.3.2 Code Cases

Several standards that are commonly invoked as code utilize Code Cases to provide alternative rules concerning materials, design, fabrication, or in-service inspection activities not covered by existing standard rules. Once approved by the SDO, code cases are effective immediately and do not expire. Code case are intended to be temporary and may be revised over time, integrated into a standard during its revision, or annulled (Fig. 1.3).

### 1.3.3 Organization of a Standard

Most standards have a consistent organizational format that allows a user to quickly navigate each standard. A standard may contain many different elements, and their format and content vary amongst the differing SDOs. The following a summary of elements that may be contained in a standard:

CASE
**2866**

ASME BPVC.CC.BPV-2023

**Approval Date: September 30, 2016**

*Code Cases will remain available for use until annulled by the applicable Standards Committee.*

**Case 2866 Pin**
**Brazing**
**Section IX**

*Inquiry:* Under what conditions may a brazing proce- dure specification (BPS) be qualified using pin brazing for attaching copper/copper alloy electrical connectors to carbon steel P-No. 101 (limited to those designated P-No. 1 for welding) pressure-retaining piping and components for applications such as cathodic protection, grounding, and nonstructural electrical connections?

*Reply:* It is the opinion of the Committee that a BPS may be qualified for pin brazing[1] attachments, provided the following conditions are met:

*(a) Brazing Procedure Specification (BPS) Qualification*

*(1)* The BPS shall be qualified by performance of ten pin braze tests using the same equipment, connector type (size, shape, and material), and position as will be used in production. The pin brazing equipment shall be complete- ly automatic except for starting. Any change to the listed variables requires requalification.

*(2)* Each of the ten brazed connections shall be tested by bending or hammering until the original brazed connection has been bent by at least 45 deg.

*(3)* In order to pass the test(s), each of the ten brazed connections shall show no more than 50% visible separation or fracture after bending.

*(b) Brazing Operator Performance Qualification.* Each brazing operator who prepares acceptable BPS qualifications is thereby qualified under the essential variables of QB-351.2. Alternatively, each brazing operator shall successfully pin braze five samples meeting the acceptance criteria for BPS qualification.

*(c)* This Case number shall be identified in the Manufacturer's Data Report.

**Fig. 1.3** Example code case for ASME Section IX

- Title Page—Descriptive title, number, identifier, etc. of the specification, date of last effective revision and revision designation
- Abstract—a summary of the standard used to help the reader quickly ascertain the standards purpose
- Legal Statements—a declaration of matters of fact including Statement of Use, Copyrights, Trademarks, ownership, and origin
- Personnel—a list of committee member, volunteers, or committees who are/where involved with the development and maintenance of the standard
- Acknowledgements or Dedication—a proclamation of gratitude for help in the creation of the standard, recognizing individuals or organizations who contributed or had an influence to the development of the standards content
- Table of Contents—a list of a standards section/clause/article titles or brief descriptions with their commencing page numbers
- List of Tables, Figures, and Forms
- Foreword, Preface, and Prologue—a brief introduction providing history and context for the standard.
- Introduction or Scope—provides an abstract of the standard's intended use, context, and organization
- Normative References—references to other standards whose requirements are mandatory
- Informative References—references to other standards whose requirements are not mandatory but provide a broader, more complete understanding of required processes and documentation
- Terms and Definitions—used to clarify what is meant by phrases and words used frequently throughout the standard. Definitions also help accurately communicate concepts to ensure consistent understanding among those applying the standard
- Body—The body contains the rules to follow when applying the standard and may be divided into sections and subsections, clauses and subclauses, and/or paragraphs. Language in the body is normative and often includes words like shall, should, may, and can indicate required items, permissible actions, and statements of possibility when conforming to the standard. Tables, Figures, and Diagrams are often included in the body to help illustrate how the standard relates to its normative references and allow users to understand the content more quickly
- Annexes and Appendices—provides additional information or details to help understand the standard. The main difference between the body and annex sections is that the body uses primarily normative, or prescriptive, language while the annex uses descriptive language
- References or Citations—provides enough bibliographic information for the reader to be able to identify and, if necessary, obtain the original resource of the information
- Synopsis or Revision Record—summarizes the chronological development or revision history.

## 1.4 Welding Standards Development Organizations (SDOs)

There are many SDOs for welding based standards. Some common U.S. based welding related SDOs include:

- ABS (American Bureau of Shipping)
- AISC (American Institute of Steel Construction)
- API (American Petroleum Institute)
- AREA (American Railway engineering Association)
- ASASHTO (American Association of State Highway and Transportation Officials)
- ASME (American Society of Mechanical Engineers)
- AWS (American Welding Society)
- AWWA (American Water Works Association)
- DOD (U.S. Department of Defense)
- NBBPVI (National Board of Boiler and Pressure Vessel Inspectors)
- NACE (National Association of Corrosion Engineers)
- SAE (Society of Automotive Engineers).

Some common Non-US welding SDOs include:

- AFNOR (Association Francaise de Normalisation)
- AS/NZS (Australian/New Zealand Standards)
- BS (British Standards)
- CEN (European Union)
- CSA (Canadian Standards Association)
- DIN (Deutsches Institute fuer Normung)
- ISO (International Organization for Standardization)
- JIS (Japanese Standards Association).

### 1.4.1 American Welding Society (AWS)

The American Welding Society (AWS) is a nonprofit organization founded in 1919 to advance the science, technology and application of welding and allied joining and cutting processes. AWS is led by a volunteer organization of officers and directors, AWS serves over 70,000 members worldwide, composed of 22 Districts with 250 Sections and student chapters. In total, AWS oversees the work of 166 committees, subcommittees, and advisory groups. (Some broader committees are subdivided into smaller, specialized subcommittees to task workload). Committees can number anywhere from 5–60 volunteers

(Average committee size is usually about 15–20 members). The main technical activities committee (TAC) oversees the activities of 29 technical committees which publish standards.

AWS maintains over 125 welding based standards which categorized as follows:

- Construction and Manufacturing (~30)
  - Structural (~7)
  - Machinery, Railroad, and Construction Equipment (~8)
  - Automotive (~3)
  - Marine (~3)
  - Piping and Tubing (~9)
  - Aerospace (~1).
- Filler materials (~33)
- Inspection: Visual (~1) and NDT (~2)
- Mechanical Testing (~2)
- Processes; Welding (~11), Brazing (~6) and Cutting (~5)
- Qualification (~11)
- Surfacing Processes (~3)
- Welding Procedures (~87).

AWS alphabetically identifies their standards by their function or intent as follows:

- A for Fundamental Technology and Consumables
- B for Qualification and Inspection
- C for Processes
- D for Industrial Applications
- E or EG for Education
- F for Safety and Health
- G for Material Weldability
- J for Welding Equipment
- QC for Certification
- Z for Safety.

### 1.4.2 American Society of Mechanical Engineers (ASME)

The American Society of Mechanical Engineers (ASME) is a nonprofit organization founded in 1880 to provide a setting for engineers to discuss the concerns brought by the rise of industrialization and mechanization. ASME later refined its mission to ensure the safety of equipment used in manufacturing and construction, particularly boilers and pressure vessels. ASME is led by Board of Governors chosen from the Society's membership who serves over 100,000 members worldwide in 140 countries.

ASME maintains over 500 standards with 5,900 volunteers on 700+committees. Its most widely known standards are its Boiler and Pressure Vessel Code (BPVC), which regulates the design and construction of boilers and pressure vessel. The BPVC, first established in 1915, covers industrial and residential boilers, nuclear reactor components, transport tanks, and other forms of pressure vessels. It consists of 12 sections with 14,000 pages, is maintained by 1,000 volunteer committee members, with 100,000 copies in use in 100 countries worldwide.

ASME also maintains its Code for Pressure Piping or B31 standards which covers the design, materials, fabrication, assembly, erection, examination, inspection, testing, operation, and maintenance of pressure piping systems. The 11 B31 standards apply to Power Piping, Fuel Gas Piping, Process Piping, Pipeline Transportation Systems for Liquid Hydrocarbons and Other Liquids, Refrigeration Piping and Heat Transfer Components and Building Services Piping.

### 1.4.3 International Organization for Standardization (ISO)

The International Organization for Standardization or ISO was founded in 1947 to facilitate and support national and international trade and commerce by developing standards that are universally recognized. ISO achieves this purpose through the participation and support of its international members who come from 164 national standards bodies to serve on its numerous technical committees. Currently, ISO has 250 technical committees, 510 subcommittees, and 2478 working groups. Since 1947, ISO has developed over 18,600 standards in every conceivable business and technical sector including energy, manufacturing, engineering, agriculture, construction, computing, metrology, healthcare, transportation, distribution, and communications.

ISO standards help to ensure that products, services, systems, and technologies work properly and are safe and effective. And to help ensure that organizations actually apply these standards, ISO's technical experts have also developed several conformity assessment guides. These guides help organizations to verify that supplies, materials, products, processes, services, systems, tools, equipment, and personnel actually comply with ISO's standards.

## 1.5 Standards Accreditation

Accreditation signifies that the process and rules used by the standard development organizations (SDOs) for the development of their standards meet the following essential requirements:

- Openness: Any party (person, organization, government) with a direct or material interest has a right to participate. Timely and adequate notice of any action to create, revise, reaffirm, or withdraw a standard shall be given to all known directly or materially affected interests.
- Balance: Interests shall not be dominated by any single interest category
- Consensus must be reached by representatives from materially affected and interested parties
- Standards are required to undergo public reviews where any member of the public may submit comments
- Comments from the consensus body and public review commenters must be responded to in good faith. An appeals process is required.

Accredited standards that meet these criteria are considered Consensus Standards. Oversight of the accreditation process is performed geographically at the national, regional, and international levels.

National Accreditation

In general, each country or economy has a single recognized National Standards Body (NSB). NSBs may be either public or private sector organizations, or combinations of the two. In some cases, NSBs also act as SDOs responsible for the lifecycle of certain standards. In the United States, the American National Standards Institute (ANSI), a non-profit organization with members from both the private and public sectors oversees the development of voluntary consensus standards for products, services, processes, systems, and personnel in the United States. ANSI also cooperates with The National Institute of Standards and Technology (NIST), the U.S. government's standards agency. A standard that has been accredited by ANSI is considered an American National Standards (ANS), which must be reviewed and revised or reaffirmed without changes at least every five (5) years. ANSI also coordinates U.S. standards with international standards so that American products can be used worldwide.

An example of an American National Standard is the American Welding Society D1.1 "Structural Welding Code—Steel" (shown in Fig. 1.4).

Regional Accreditation

Regional and sub-regional standards organizations also exist which bring together regional NSBs to work collectively on standards of interest to a geographical region. Like NSBs,

AWS D1.1/D1.1M:2020
An American National Standard

Approved by the
American National Standards Institute
December 9, 2019

**Structural Welding Code—Steel**

**Statement on the Use of American Welding Society Standards**

All standards (codes, specifications, recommended practices, methods, classifications, and guides) of the American Welding Society (AWS) are voluntary consensus standards that have been developed in accordance with the rules of the American National Standards Institute (ANSI). When AWS American National Standards are either incorporated in, or made part of, documents that are included in federal or state laws and regulations, or the regulations of other governmen- tal bodies, their provisions carry the full legal authority of the statute. In such cases, any changes in those AWS standards must be approved by the governmental body having statutory jurisdiction before they can become a part of those laws and regulations. In all cases, these standards carry the full legal authority of the contract or other document that invokes the AWS standards. Where this contractual relationship exists, changes in or deviations from requirements of an AWS stand- ard must be by agreement between the contracting parties.

AWS American National Standards are developed through a consensus standards development process that brings together volunteers representing varied viewpoints and interests to achieve consensus. While AWS administers the proc- ess and establishes rules to promote fairness in the development of consensus, it does not independently test, evaluate, or verify the accuracy of any information or the soundness of any judgments contained in its standards.

**Fig. 1.4** Example frontmatter for AWS D1.1 showing national accreditation

regional standards organizations also act as SDOs responsible for the lifecycle of certain standards. For example, in the European Union, the European Committee for Standardization (CEN) is an association that brings together the NSBs of 34 European countries and is responsible for developing and defining voluntary standards at European level. A standard that has been accredited by CEN is considered a European Standard (EN) and is typically published in many different languages of its member countries.

An example of a European Standard is ISO 15612 "Specification and qualification of welding procedures for metallic materials—Qualification by adoption of a standard welding procedure" (shown in Fig. 1.5).

International Accreditation

An international standards organization develops and/or accredits standards on a global or international level. While there are several international standards organizations, the International Organization for Standardization (ISO) handles welding related standards. ISO is an independent non-governmental organization composed of the NSBs from 165 member countries whose goal is to facilitate the international coordination and unification of industrial standards. ISO has developed over 23553 International Standards.

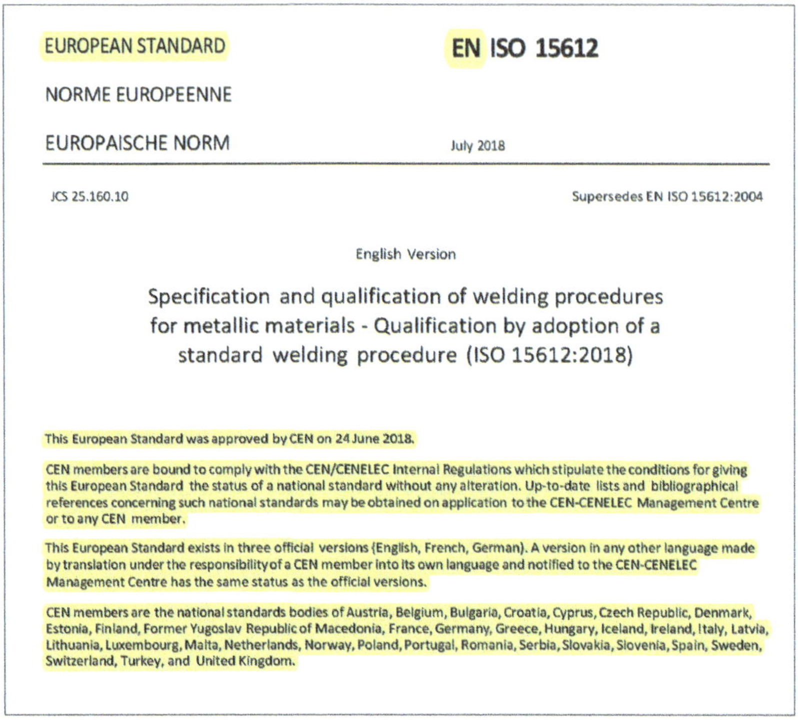

EUROPEAN STANDARD

NORME EUROPEENNE

EUROPAISCHE NORM

**EN ISO 15612**

July 2018

ICS 25.160.10

Supersedes EN ISO 15612:2004

English Version

Specification and qualification of welding procedures for metallic materials - Qualification by adoption of a standard welding procedure (ISO 15612:2018)

This European Standard was approved by CEN on 24 June 2018.

CEN members are bound to comply with the CEN/CENELEC Internal Regulations which stipulate the conditions for giving this European Standard the status of a national standard without any alteration. Up-to-date lists and bibliographical references concerning such national standards may be obtained on application to the CEN-CENELEC Management Centre or to any CEN member.

This European Standard exists in three official versions (English, French, German). A version in any other language made by translation under the responsibility of a CEN member into its own language and notified to the CEN-CENELEC Management Centre has the same status as the official versions.

CEN members are the national standards bodies of Austria, Belgium, Bulgaria, Croatia, Cyprus, Czech Republic, Denmark, Estonia, Finland, Former Yugoslav Republic of Macedonia, France, Germany, Greece, Hungary, Iceland, Ireland, Italy, Latvia, Lithuania, Luxembourg, Malta, Netherlands, Norway, Poland, Portugal, Romania, Serbia, Slovakia, Slovenia, Spain, Sweden, Switzerland, Turkey, and United Kingdom.

**Fig. 1.5** Example frontmatter for ISO 15612 showing regional accreditation

ISOs TC44 is the main technical committee on "Welding and Allied Processes", which consists of 10 subcommittees. Their scope is the standardization of welding and allied processes including terminology, definitions, and the symbolic representation of welds on drawings, apparatus and equipment for welding, raw materials (gas, parent, and filler metals) welding processes and rules, methods of test and control, calculations and design of welded assemblies, welders' qualifications, as well as safety and health. To-date, ISO has developed ~120 international standards related to welding and allied processes.

An example of an ISO accredited welding standard is ISO 15614-1 "Specification and qualification of welding procedures for metallic materials—Welding procedure test—Part 1: Arc and gas welding of steels and arc welding of nickel and nickel alloys" (shown in Fig. 1.6).

INTERNATIONAL STANDARD

ISO 15614-1

Second edition
2017-06

**Specification and qualification of welding procedures for metallic materials — Welding procedure test —**

Part 1:

**Arc and gas welding of steels and arc welding of nickel and nickel alloys**

**Foreword**

ISO (the International Organization for Standardization) is a worldwide federation of national standards bodies (ISO member bodies). The work of preparing International Standards is normally carried out through ISO technical committees. Each member body interested in a subject for which a technical committee has been established has the right to be represented on that committee. International organizations, governmental and non-governmental, in liaison with ISO, also take part in the work. ISO collaborates closely with the International Electrotechnical Commission (IEC) on all matters of electrotechnical standardization.

**Fig. 1.6** Example frontmatter for ISO 15614-1 showing international accreditation

## 1.6 Standards Compliance

In general standards can classified as:

- "de Facto" (or "Ad Hoc") standards which means they are followed by informal convention or widely adopted by an industry and its customers but have not been adopted by a Standards Development Organization (SDO). Its use is strictly voluntary.
- "Formal" standards are formally developed and published by an SDO. Its use is voluntary unless invoked by legally binding contracts, laws or regulations which then classifies it as a "de Jure" standard.

- "de Jure" standards are part of legally binding contracts, laws, or regulations, and are endorsed by a formal Standards Development Organization (SDO) or regulatory agency. Its use is mandatory.

A "Code" is a specific type of "de Jure" standard that has been enacted into law by a local, regional, or national authority having jurisdiction so that the engineer or contractor is legally obligated to comply with the code. Noncompliance can result in being prosecuted. The code may be an industry, government, or voluntary consensus-based standard. A code can include references to standards, which means the standards are incorporated by reference and therefore are part of the code and legally enforceable.

Examples of de Jure welding standards commonly invoked as Code include:

- AWS D1.1 "Structural Welding Code—Steel"
- ASME "Boiler and Pressure Vessel Code".

Standards themselves vary in their method of achieving compliance. Some have specific requirements that do not allow for alternative actions. Others permit alternative actions or procedures, so long as they result in properties that meet specified criteria. These criteria are often given as minimum requirements.

> Example: "Ultimate tensile strength of welded specimen must meet or exceed the minimum tensile strength specified for the base metal being welded"

Standards frequently use auxiliary verbs to signify requirements. The choice of an auxiliary verb in a standard affects the significance of a requirement. Auxiliary verbs are typically capitalized and may include the following:

- SHALL, MUST, WILL or REQUIRED—Establishes a mandatory minimum requirement or practice.
- SHOULD or RECOMMENDED—Establishes a recommended requirement or practice.
- MAY or OPTIONAL—Establishes an optional requirement or practice.

Examples of the use of auxiliary verbs in the different types of standards include:

- Codes typically use SHALL, MUST, or REQUIRED
- Specifications typically use "SHALL"
- Guides typically use "SHOULD" or "MAY"
- Methods typically use "SHOULD" or "MAY"
- Recommended Practices typically use "SHOULD" or "MAY".

Additional keywords of importance in standards include:

- "typical," "for example," "for reference," or the Latin abbr. "e.g."—Indicates suggested practices given for guidance only.
- "or" used in conjunction with a mandatory requirement or a recommended practice indicates that there are two or more options for complying with the stated requirement or practice.

## 1.7 Summary

Familiarity with a standards' organization and knowledge of its contents is important in order to be proficient in its use. Standards should not be memorized because their requirements are constantly changing.

There are many factors to consider when applying standards including its:

- Scope and intended application
- Use the index, table of contents, etc.
- Mandatory versus Non-Mandatory Requirements and Recommendations
- Interpretation and intent of its Requirements and Recommendations
- Normative versus Informative References.

**Notes**

1. 2005, Stacy Finz, "Hosed/S.F. hydrants don't fit equipment from other fire departments. In a disaster, the city could be …", San Francisco Chronicle. (https://www.sfgate.com/news/article/HOSED-S-F-hydrants-don-t-fit-equipment-from-2568046.php).
2. 2016, Circular A-119 Fed Participation in the Dev & Use of Voluntary Consensus Standards and in Conformity Assessment Activities, Office of Management and Budget, Executive Office of the President.

# Welding Qualification Standards

# 2

There are many objectives and benefits of qualifying welding procedures and welders/operators. Qualification to a standard provides assurance that only safe and reliable welded products are produced.

The purpose of qualifying a welding procedure is to prove that it is capable of producing welds meet the minimum required mechanical and metallurgical properties for the intended application. Qualification of the proposed welding procedure demonstrates the procedure itself will satisfy the design requirements, and not the skill of the welder or welding operator. The qualification is performed with the proposed welding procedure on a qualification test weldment which is then evaluated by both non-destructive evaluation and destructive testing. After successfully meeting the acceptance criteria, the qualification is formally documented as a Procedure Qualification Record or PQR (see Sect. 3.2 and Chap. 5), and the proposed welding procedure is finalized as a Welding Procedure Specification or WPS (see Sect. 3.1 and Chap. 5). This process is generically referred to as "procedure qualification" in this text.

The purpose of qualifying a welder or welding operator is to prove the individuals' skills have been verified and that they can follow the instructions in a given welding procedure specification. The qualification is performed with a qualified WPS on a qualification test weldment which is then evaluated by both non-destructive evaluation and destructive testing. After successfully meeting the acceptance criteria, the qualification is formally documented as a welder/operator Performance Qualification Test Record or PQTR (see Sect. 3.3 and Chap. 8). This process is generically referred to as "performance qualification" in this text.

© The Author(s), under exclusive license to Springer Nature Switzerland AG 2025

D. Barborak, *Arc Welding Qualification Standards*, Synthesis Lectures on Welding Engineering, https://doi.org/10.1007/978-3-031-64646-1_2

## 2.1 AWS B2.1 Specification for Welding Procedure and Performance Qualification

AWS B2.1 Specification for Welding Procedure and Performance Qualification, hereinafter referred to as AWS B2.1, provides the requirements for qualification of welding procedure specifications, welders, and welding operators for manual, semiautomatic, mechanized, and automatic welding. Provisions included in AWS B2.1 include base metals, filler metals, qualification variables, welding designs, and testing requirements are also included.

The first edition of AWS B2.1 was published in 1984, with subsequent revisions in 1998, 2000, 2005, 2009, 2013, and its current revision of 2021. AWS B2.1 is currently on a 5-year revision cycle.

AWS B2.1/B2.1M:2021
An American National Standard

Approved by
the American National Standards
Institute
November 26, 2021

**Specification for**
**Welding Procedure and**
**Performance Qualification**

7th Edition

Revises AWS B2.1/B2.1M:2014

Prepared by the
American Welding Society (AWS) B2 Committee on Procedure and Performance Qualification

Under the Direction of the
AWS Technical Activities Committee

Approved by the
AWS Board of Directors

**American Welding Society**

Some notable changes in its recent history include:

- 1998 Allowed qualification by either a standard test or workmanship test versus qualification testing conducted with a specific end use application in mind.
- 2000 Added qualified thickness limitations for GMAW-S, allows qualification of material M-Number M-5A or lower welded to itself and to each of the lower M-Number metals.
- 2009 Major reformatting, the deletion of any reference to "sheet metals", new welding and qualification variables and tighter restrictions.
- 2014 Heat input formulae for waveform controlled power sources, the addition of NAVSEA/MIL Standard Welding Procedure Specifications.

The organization of AWS B2.1 is divided in five main clauses as follows:

- Clause 1 General Requirements
- Clause 2 Normative References
- Clause 3 Terms and Definitions
- Clause 4 Procedure Qualification
- Clause 5 Performance Qualification.

These clauses contain general references and guides that apply to procedure and performance qualifications such as positions, type and purpose of various mechanical tests, acceptance criteria, and the applicability of AWS B2.1.

The arc welding processes covered by AWS B2.1 include:

- Shielded Metal Arc Welding (SMAW)
- Submerged Arc Welding (SAW)
- Gas Metal Arc Welding (GMAW)
- Flux Cored Arc Welding (FCAW)
- Gas Tungsten Arc Welding (GTAW)
- Plasma Arc Welding (PAW).

AWS B2.1 also include qualification requirements for other welding processes namely Oxyfuel Gas Welding (OFW), Electroslag Welding (ESW), Electrogas Welding (EGW), Electron Beam Welding (EBW), Stud Welding (SW), and Laser Beam Welding (LBW).

### 2.1.1 Clause 1 General Requirements

Clause 1 covers the scope of AWS B2.1 including welding processes, responsibility, units of measure, and safety.

### 2.1.2 Clause 2 Normative References

Clause 2 references the following standards which contain mandatory provisions referenced throughout AWS B2.1:

- AWS A2.4 *Standard Symbols for Welding, Brazing, and Nondestructive Examination*
- AWS A3.0 *Standard Welding Terms and Definitions, Including Terms for Adhesive Bonding, Brazing, Soldering, Thermal Cutting, and Thermal Spraying*
- AWS B4.0 *Standard Methods for Mechanical Testing of Welds*
- AWS D1.4 *Structural Welding Code—Reinforcing Steel*
- AWS D11.2 *Guide for Welding Iron Casting*
- ANSI Z49.1 *Safety in Welding, Cutting, and Allied Processes*
- ASME BPVC V *Nondestructive Examination*
- ASME BPVC IX *Welding, Brazing, and Fusing Qualifications*
- ASTM E165 *Standard Test Method for Liquid Penetrant Examination.*

### 2.1.3 Clause 3 Terms and Definitions

Clause 3 identifies terms and definitions specifically used in AWS B2.1 in addition to the reference to AWS A3.0 "Standard Welding Terms and Definitions" called out in AWS B2.1 Clause 2 Normative References.

### 2.1.4 Clause 4 Procedure Qualification

Clause 4 covers the rules for qualification of a welding procedure and preparation of the PQR and WPS documentation. General provisions are given in subclause 4.1 and the user is given the requirements for several options for a qualified procedure, qualification by a standard test weldment, qualification by a special test weldment (simulated service or prototype) or use of a Standard Welding Procedure Specification SWPS (subclause 4.2).

Subclause 4.3 outlines the requirements for procedures qualified by the employer. PQRs shall not be revised except to correct errors or add new or omitted information. All such changes shall be identified, authorized by the Employer, and dated on the PQR. A WPS may require the support of more than one PQR, while one PQR may support several WPSs. Compatible PQRs using different qualification variables may be combined to support one WPS. Compatibility shall be determined based on either experience, or testing, or both. WPSs and PQRs shall be identified according to a system that allows permanent traceability from the WPS to its supporting PQRs. A change in a WPS beyond that allowed in subclause 4.14, Procedure Qualification Variables shall require requalification of the procedure and preparation of a new or revised WPS. Changes not addressed in subclause 4.14 shall not require requalification, provided such changes are documented in a new or revised WPS. WPSs qualified to earlier editions of AWS B2.1 shall be qualified to this edition without revision or further testing.

The qualification thickness limitations are outlined in subclause 4.5 and Tables 4.2 through 4.5.

The qualification test requirements and acceptance criteria are separately provided for special test weldments (subclause 4.6), pipe and plate groove test weldments (subclause 4.7), fillet welds (subclause 4.8), cladding welds (subclause 4.9), hardfacing welds (subclause 4.10), stud welds (subclause 4.11), and test weldments less than 1.5 mm (1/16 in.) thick (subclause 4.11).

The essential, supplementary essential, and non-essential welding variables are addressed separately for each welding process in subclauses 4.13 and 4.14.

### 2.1.5 Clause 5 Performance Qualification

Clause 5 covers the requirements for welder and welding operator performance qualification. The qualification can be performed on a Standard Test Weldment (subclause 5.2) or Workmanship Test Weldment (subclause 5.3). Clause 5 covers responsibility, type of tests, records, welder identification, positions, diameters, expiration and renewal of qualifications. Welders and Welding Operators may be qualified by visual and mechanical tests, or by radiography of a test coupon, or by radiography of the initial production weld.

The Performance Qualification Test Record (PQTR) shall include the following:

- Identify the WPS or SWPS utilized
- Address each of the qualification variables outlined in subclause 5.6
- Identify the test and examination methods utilized and their results based on the requirements outlined in Table 5.1 and Subclause 5.5
- Identify the limits of qualification for the welder or welding operator outlined in subclause 5.6.

### 2.1.6 Annexes

AWS B2.1 contains the following annexes:

- Annex A (Normative)—Illustrations—Weld Position, Test Specimens, and Text Fixtures
- Annex B (Normative)—Filler Metal Grouping
- Annex C (Normative)—Base Metal Grouping
- Annex D (Normative)—Radiographic Examination Procedure
- Annex E (Informative)—Basis for Establishing a Standard Welding Procedure Specification
- Annex F (Informative)—Sample Forms
- Annex G (Informative)—Macroetch Procedure
- Annex H (Informative)—Guidelines for the Preparation of Technical Inquiries
- Annex I (Informative)—Guide for Requesting Adoption of New Materials Under the AWS B2.1.

### 2.1.7 Other AWS Qualification Standards

Prior to the adoption of AWS B2.1 in 1984, many of the AWS construction/fabrication standards had their own separate qualification requirements. Since then, the revisions of most of these standards either partially or fully recognize or invoke the requirements of AWS B2.1 for qualification. Below is a summary of some of these standards.

AWS standards that invoke AWS B2.1 for qualification requirements:

- D1.6 *Structural Welding Code—Stainless Steel*
- D9.1 *Sheet Metal Welding Code*
- D14.3 *Welding Earthmoving Construction and Agricultural Equipment*
- D14.4 *Design of Welded Joints in Machinery and Equipment*
- D14.5 *Welding of Presses and Press Components*

- D14.6 *Welding of Rotating Elements*
- D17.1 *Fusion Welding for Aerospace Applications.*

AWS standards that accept procedures qualified under AWS B2.1 with Engineer's approval:

- D1.1 *Structural Welding Code—Steel*
- D1.2 *Structural Welding Code—Steel*
- D1.3 *Structural Welding Code—Sheet Steel*
- D1.8 *Structural Welding Code—Seismic Supplement*
- D1.9 *Structural Welding Code—Titanium.*

AWS standards that accept procedures qualified under AWS B2.1 only if they meet additional requirements:

- D1.4 *Structural Welding Code—Reinforcing Steel*
- D1.5 *Bridge Welding Code*
- D3.6 *Underwater Welding Code*
- D14.1 *Welding of Industrial and Mill Cranes and Other Material Handling Equipment*
- D14.3 *Welding Earthmoving Construction and Agricultural Equipment*
- D14.7 *Surfacing and Reconditioning of Industrial Mill Rolls*
- D14.9 *Welding of Hydraulic Cylinders*
- D15.1 *Railroad Welding Spec for Cars and Locomotives*
- D15.2 *Welding of Rails and Related. Rail Components for Use by Rail Vehicles*
- D18.1 *Welding Of Austenitic Stainless Steel Tube And Pipe Systems In Sanitary Hygienic Applications*
- D18.3 *Welding of tanks Vessels and Other Equipment in Sanitary Applications.*

## 2.2 ASME BPVC ASME IX Welding, Brazing, and Fusing Qualifications

ASME IX of the ASME Boiler and Pressure Vessel Code, hereinafter referred to as ASME IX, relates to the qualification of welders, welding operators, brazers, brazing operators, and fusing operators, and the procedures employed in welding, brazing, or plastic fusing. ASME IX does not contain rules for production joining, nor does it contain rules to cover all factors affecting production material joining properties under all circumstances.

The first edition of ASME IX was first published in 1941 to consolidate the existing welding qualification requirements in the construction codes. ASME IX has been revised numerous times since its inception and currently on a 2-year revision cycle.

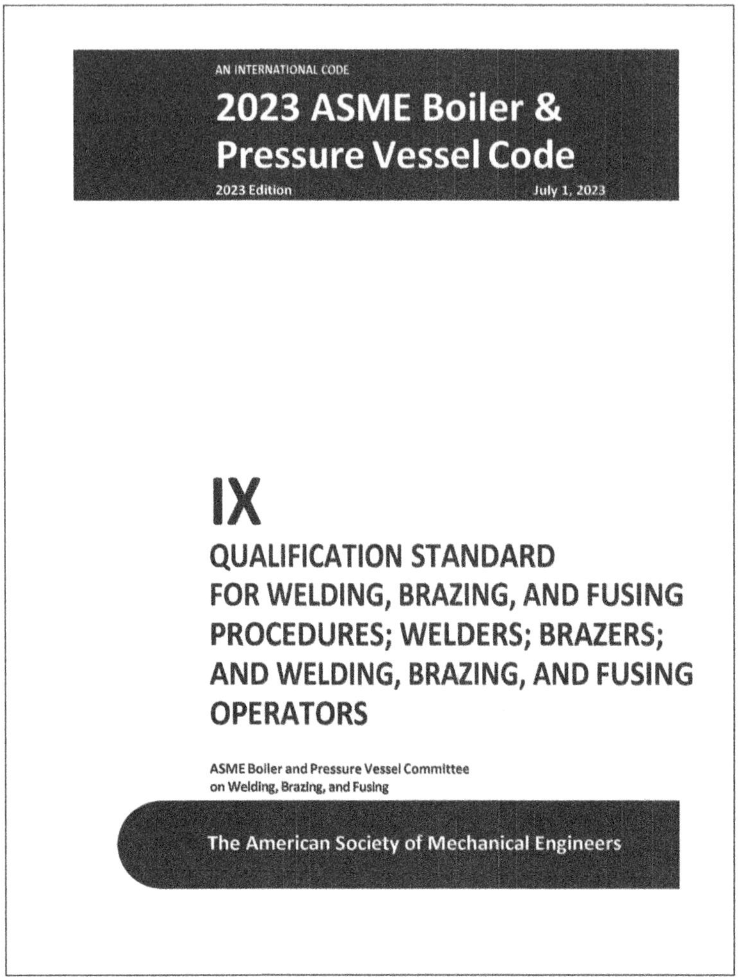


Some notable changes in its recent history include:

- 1974 The rules for notch-toughness were separated from the essential variables as supplementary essential variables.
- 1980 New definitions for position were added along with their corresponding qualification.
- 1994 Welding P-Numbers, brazing P-Numbers, and nonmandatory S-Numbers were consolidated into one table. Metals were listed in numerical order by material specification number to aid users in locating the appropriate grouping number.
- 2000 The use of Standard Welding Procedure Specifications (SWPSs) was permitted.
- 2004 Rules for temper bead welding were added.

- 2009 The S-Number base metals were reassigned as P-Numbers and the S-Number listings and references were deleted.
- 2010 Rules for the measurement of heat input for waveform controlled welding power supplies were added.
- 2013 Rules for Plastic Fusion were added.

ASME IX is divided into four Parts, three of which cover specific material joining process requirements for Welding, Brazing, and Plastic Fusing. Each Part addressing a material joining process is further divided into Articles as follows:

- Part QG—General Requirements
- Part QW—Welding
  - Article I: Welding General Requirements
  - Article II: Welding Procedure Qualifications
  - Article III: Welding Performance Qualifications
  - Article IV: Welding Data
  - Article V: Standard Welding Procedure Specifications.
- Part QB—Brazing
  - Article XI: Brazing General Requirements
  - Article XII: Brazing Procedure Qualifications
  - Article XIII: Brazing Performance Qualifications
  - Article XIV: Brazing Data.
- Part QF—Plastic Fusing
  - Article XXI: Plastic Fusing General Requirements
  - Article XXII: Fusion Procedure Qualifications.

These articles contain general references and guides that apply to procedure and performance qualifications such as positions, type and purpose of various mechanical tests, acceptance criteria, and the applicability of ASME IX. The general requirement articles reference the data articles for specific details of the testing equipment and removal of the mechanical test specimens.

The arc welding processes covered by ASME IX Part QW include:

- Shielded Metal Arc Welding (SMAW)
- Submerged Arc Welding (SAW)
- Gas Metal Arc Welding (GMAW)
- Flux Cored Arc Welding (FCAW)
- Gas Tungsten Arc Welding (GTAW)
- Plasma Arc Welding (PAW).

ASME IX Part QW also include qualification requirements for other welding processes namely Oxyfuel Gas Welding (OFW), Electroslag Welding (ESW), Electrogas Welding (EGW), Electron Beam Welding (EBW), Stud Welding (SW), Inertia and Continuous Drive Friction Welding (FRW), Resistance Welding (RW), Laser Beam Welding (LBW), Lower Power Density Laser Beam Welding (LLBW), Flash Welding (FW), Diffusion Welding (DFW), and Friction Stir Welding (FSW).

### 2.2.1 Part QW Article I—Welding General Requirements QW-100

Article I covers the scope of ASME IX, the purpose and use of the WPS, PQR and WPQ, responsibility, test positions, types and purposes of tests and examinations, test procedures, acceptance criteria, visual examination, and radiographic examination.

### 2.2.2 Part QW Article II—Welding Procedure Qualifications QW-200

Article II covers the rules for the preparation of WPS and PQR. Each welding process is listed separately in QW-250 with the essential, supplementary essential, and nonessential variables.

The WPS shall specify a value or range for each essential, nonessential and, when necessary, each supplementary essential variable listed for each welding process. The PQR shall record the value for each essential and, when necessary, each supplementary essential variable used.

When a change is made in an essential variable, the WPS must be revised, and requalified with a new PQR, unless the revision can be supported by an existing PQR.

Similarly, when the code requires notch-toughness, supplementary essential variables become additional essential variables. When a change is made in a supplementary essential variable, the WPS must be revised, and requalified by a new PQR, unless the revision can be supported by an existing PQR for notch-toughness applications.

### 2.2.3 Part QW Article III—Welding Performance Qualifications QW-300

Article III covers the preparation of WPQ records. Each welding process is listed separately in QW-350 with the essential variables for welding performance. The WPQ form must record a value for each essential variable used and must list a range qualified for each of these essential variables.

Article III covers responsibility, type of tests, records, welder identification, positions, diameters, expiration and renewal of qualifications. Welders and Welding Operators may be qualified by visual and mechanical tests, or by radiography of a test coupon, or by radiography of the initial production weld.

### 2.2.4 Part QW Article IV—Welding Data QW-400

Article IV covers welding variables that are used in the preparation and qualification of the WPS, PQR or the WPQ as applicable. For convenience the welding variables are grouped into the following categories:

- Joints
- Base Materials
- Filler Materials
- Positions
- Preheat
- Post-Weld Heat Treatment
- Gas
- Electrical Characteristics
- Technique.

Article IV also includes assignments of P-Numbers (ASME base materials), F-Numbers (grouping of filler metals) and A-Numbers (weld metal chemical analysis).

Tables for WPS qualification thickness limits and tables for WPQ thickness and diameter limits are also included. There are tables for welding positions signifying that a welder who qualifies in a particular position is qualified to weld within a range of positions as appropriate. Furthermore, test coupons, the removal of test specimens, and the test jig dimensions are identified.

### 2.2.5 Part QW Article V—Standard Welding Procedure Specifications (SWPS) QW-500

Article V covers rules for the adoption, demonstration, and application of the Standard Welding Procedure Specifications, (SWPSs).

### 2.2.6 Appendices

ASME IX contains the following appendices:

- Nonmandatory Appendix B: Welding and Brazing Forms
- Nonmandatory Appendix D: P-Number Listing
- Mandatory Appendix E: Permitted SWPSs
- Mandatory Appendix F: Standard Units for Use in Equations
- Nonmandatory Appendix G: Guidance for the Use of U.S. Customary and SI Units in the ASME Boiler and Pressure Vessel Code
- Nonmandatory Appendix H: Waveform Controlled Welding
- Mandatory Appendix J: Guideline for Requesting P-Number Assignments for Base Metals not Listed in Table QW/QB-422
- Nonmandatory Appendix K: Guidance on Invoking ASME IX Requirements in Other Codes, Standards, Specifications, and Contract Documents
- Nonmandatory Appendix L: Welders and Welding Operators Qualified Under ISO 9606–1:2012 and ISO 14732–2013.

### 2.2.7 Other ASME Qualification Standards

Since the introduction of ASME IX in 1941, most of the ASME construction/fabrication standards invoke the requirements of ASME IX for qualification. Below is a summary of some of these standards.

ASME standards that invoke ASME IX for qualification requirements:

- BPVC.I *Rules for Construction of Power Boilers*
- BPVC.III *Rules for Constructions of Nuclear Facility Components*
- BPVC.IV *Rules for Construction of Heating Boilers*
- BPVC.VIII *Rules for Construction of Pressure Vessels*
- BPVC.XI *Rules for In-service Inspection of Nuclear Power Plant Components*
- B31.1 *Power Piping*
- B31.3 *Process Piping*
- B31.8 *Gas Transmission and Distribution Piping Systems*
- PCC-2 *Repair of Pressure Equipment and Piping*.

ASME standards that conditionally invoke ASME IX for qualification requirements:

- B31.4 *Liquid Transportation Systems for Hydrocarbons, Liquid Petroleum Gas, Anhydrous Ammonia and Alcohols*
- B31.5 *Refrigeration Piping*
- B31.9 *Building Services Piping*
- B31.12 *Hydrogen Piping*.

## 2.3 Other Welding Qualification Standards

Most of the fabrication and construction standards from various industries either contain their own requirements for welding procedure and personnel qualification or reference a separate standalone qualification standard. The following qualification standards for ISO and CSA may also be applicable in North America.

### 2.3.1 International Standards Organization

While AWS and ASME generally provides all procedure and welder/operator qualification requirements into a single document, ISO specifies qualification requirements in a series of separate documents.

The following ISO standards are required for welding procedure qualification:

- ISO 15607 *Specification and Qualification of Welding Procedures for Metallic Materials—General Rules* defines general rules for the specification and qualification of welding procedures for metallic materials. It also refers to several other standards as regards detailed rules for specific applications and is applicable to manual, mechanized and automatic welding.
- ISO/TR 15608 *Welding—Guidelines for a Metallic Materials Grouping System* provides guidelines for a uniform system for grouping materials for welding purposes. It can also be applied for other purposes, such as heat treatment, forming and non-destructive testing.
- ISO 15609-1 *Specification and Qualification of Welding Procedures for Metallic Materials—Welding Procedure Specification—Part 1: Arc Welding* specifies requirements for the content of welding procedure specifications for arc welding processes. The variables listed are those influencing the quality of the welded joint.
- ISO 15610 *Specification and Qualification of Welding Procedures for Metallic Materials—Qualification Based on Tested Welding Consumables* explains the requirements for the qualification of welding procedures based on tested consumables and specifies the range of qualification. This standard is not applicable where requirements for hardness or impact properties, preheating, controlled heat input, inter-pass temperature and post-weld heat-treatment are specified for the welded joint.
- ISO 15611 *Specification and Qualification of Welding Procedures for Metallic Materials—Qualification Based on Previous Welding Experience* explains the requirements for the qualification of welding procedures based on previous welding experience and specifies the range of qualification.
- ISO 15612 *Specification and Qualification of Welding Procedures for Metallic Materials—Qualification by Adoption of a Standard Welding Procedure* explains the requirements for adoption of a standard welding procedure, and establishes the conditions,

limits and ranges of qualification necessary for the use of a standard welding procedure. In addition, it gives the manufacturer the possibility of using welding procedures based on welding procedure tests performed by other organizations.

- ISO 15613 *Specification and Qualification of Welding Procedures for Metallic Materials—Qualification Based on Pre-production Welding Test* specifies how a preliminary welding procedure specification is qualified based on pre-production welding tests.
- ISO 15614 *Specification and Qualification of Welding Procedures for Metallic Materials—Welding Procedure Test* is a series of standards that specifies how a preliminary welding procedure specification is qualified by welding procedure tests. This series includes the following standards broken out by material type and application:
  - ISO 15614-1 *Part 1: Arc and Gas Welding of Steels and Arc Welding of Nickel and Nickel Alloys*
  - ISO 15614-2 *Part 2: Arc Welding of Aluminum and its Alloys*
  - ISO 15614-3 *Part 3: Fusion Welding of Non-alloyed and Low-alloyed Cast Irons*
  - ISO 15614-4 *Part 4: Finishing Welding of Aluminum Castings*
  - ISO 15614-5 *Part 5: Arc Welding of Titanium and Zirconium and their Alloys*
  - ISO 15614-6 *Part 6: Arc and Gas Welding of Copper and its Alloys*
  - ISO 15614-7 *Part 7: Overlay Welding*
  - ISO 15614-8 *Part 8: Welding of Tubes to Tube-plate Joints.*

The following ISO standards are required for welder/operator qualification:

- ISO 9606 *Qualification Testing of Welders—Fusion Welding* is a series of standards that specifies the requirements for qualification testing of welders for fusion welding. They provide a set of technical rules for a systematic qualification test of the welder, and enables such qualifications to be uniformly accepted independently of the type of product, location and examiner or examining body. When qualifying welders, the emphasis is placed on the welder's ability manually to manipulate the electrode, welding torch or welding blowpipe, thereby producing a weld of acceptable quality. The welding processes referred to include those fusion-welding processes which are designated as manual or partly mechanized welding. It does not cover fully mechanized and automated welding processes. This series contains the following standards broken out by material type:
  - ISO 9606-1 *Part 1: Steels*
  - ISO 9606-2 *Part 2: Aluminum and Aluminum Alloys*
  - ISO 9606-3 *Part 3: Copper and Copper Alloys*
  - ISO 9606-4 *Part 4: Nickel and Nickel Alloys*
  - ISO 9606-5 *Part 5: Titanium and Titanium Alloys, Zirconium and Zirconium Alloys.*
- ISO 14732 *Welding Personnel—Qualification Testing of Welding Operators and Weld Setters for Mechanized and Automatic Welding of Metallic Materials* specifies requirements for qualification of welding operators and also weld setters for mechanized and automatic welding.

### 2.3.2 Canadian Standards Association

The Canadian SDO known as the Canadian Standards Association or CSA has several standards that contain requirements for welding procedure and personnel qualification, namely:

- CSA W47.1 *Certification of Companies for Fusion Welding of Steel*
- CSA W47.2 *Fusion Welding of Aluminum Company Certification*
- CSA W59 *Welded steel construction*
- CSA W59.2 *Welded Aluminum Construction*
- CSA W186 *Welding of Reinforcing Bars in Reinforced Concrete Construction*
- CSA Z662 *Oil and Gas Pipeline Systems.*

These standards specify the requirements for both a Welding Procedure Specification (WPS) and a Welding Procedure Data Sheet (WPDS) to provide direction to the welding supervisor, welders and welding operators. The WPS provides general information on the welding process and material grouping being welded, while the WPDS provides specific welding variables/parameter/conditions for the specific weldment. All WPS and WPDS must be independently reviewed and accepted by the Canadian Welding Bureau (CWB) prior to use. These CSA standards also define requirements for procedure qualification testing (PQT) to support the acceptance of the WPDS. A record of the procedure qualification test and the results must be documented on a procedure qualification record (PQR). Finally, these standards outline the requirements for certification of the welding supervisors, welding engineers, and companies or organizations that perform fabrication, inspection, or produce materials such as welding consumables.

By Canadian law, all welded fabrication activities must be verified and/or witnessed by the CWB. CWB certification means that a company's welding program, welding procedures, and welder qualifications and have been independently verified by a third-party that all relevant standards are followed.

## 2.4 Summary

There are numerous qualification requirements outlined in the various industry standards. Many of requirements can be interrelated and/or dependent with the requirements from other standards. Sometimes multiple qualification standards may be invoked with conflicting requirements.

Wikipedia provides a comprehensive list of Standards Development Organizations (SDOs) and their welding related standards.[1] For reference, this book will primarily draw upon the requirements of the American Welding Society Specification for Welding Procedure and Performance Qualification (AWS B2.1), and provisions of the American Society of Mechanical Engineers specification for "Qualification Standard for Welding, Brazing, and Fusing Procedures; Welders; Brazers; and Welding, Brazing and Fusing Operators" (ASME IX).

**Note**

1. https://en.wikipedia.org/wiki/List_of_welding_codes.

# 3 Welding Related Documentation

Many welding related standards as well as individual organizations quality management manuals and procedures require the organization to record and keep certain welding related documentation. The purpose of this is to communicate information and provide evidence of conformity. With respect to welding qualification, an organization should maintain a record of the results obtained in welding procedure and welder and welding operator performance qualifications (Ref. ASME IX QW-103.2).

The primary forms of welding documentation include the:

- Welding Procedure Specification (WPS)
- Procedure Qualification Record (PQR)
- Performance Qualification Test Record (PQTR).

There are also numerous forms of supporting documentation including:

- Material Test Reports (MTR)
- Non-Destructive Examination Reports
- Destructive Examination Reports
- Weld Technique Sheets
- Weld Bead Logs
- Post-weld Heat Treatment Reports.

© The Author(s), under exclusive license to Springer Nature Switzerland AG 2025

D. Barborak, *Arc Welding Qualification Standards*, Synthesis Lectures on Welding Engineering, https://doi.org/10.1007/978-3-031-64646-1_3

## 3.1 Welding Procedure Specification (WPS)

AWS B2.1 defines a WPS as a document providing the required welding variables for a specific application to assure repeatability by properly trained welders and welding operators. Alternatively, ASME IX defines a WPS as a written qualified welding procedure prepared to provide direction for making production welds to Code requirements. The WPS or other documents may be used to provide direction to the welder or welding operator to assure compliance with the Code requirements.

For convenience, the welding variables are typically grouped into the following categories on a WPS:

- Joint Details
- Base Materials
- Filler Materials
- Positions
- Preheat
- Post-Weld Heat Treatment
- Gas
- Electrical Characteristics
- Technique.

See Chap. 5 for a detailed discussion on welding variables. An example of a Welding Procedure Specification is shown in Fig. 3.1 with variable groupings highlighted in yellow.

Any format for the welding procedure specification may be used provided all applicable information is recorded (ref. AWS B2.1 4.1.3 & AWS IX QW-200.1(d)). Examples of WPS formats are provided in AWS B2.1 Annex F Form F.2, and ASME IX Nonmandatory Appendix B Form QW-482.

### 3.1.1 Preliminary Welding Procedures Specification (pWPS)

A preliminary welding procedure (pWPS) is simply a draft WPS that is utilized during the procedure qualification process to guide the welder/operator in completing the test coupon. The format of a pWPS is typically the same format utilized for the qualified WPS.

**WELDING PROCEDURE SPECIFICATION (WPS)**

| | | | | | |
|---|---|---|---|---|---|
| RED Inc. | | WPS-01-01-F-231 | 0 | 12/01/2020 | |
| Company Name | | WPS No. | Rev. No. | Date | |
| J. Jones | 12/01/2020 | PQR-01-01-F-231 | No | | |
| Authorized by | Date | Supporting PQR(s) | CVN Report | | |

| BASE METALS | Specification | Type or Grade | AWS Group No. |
|---|---|---|---|
| Base Material | ASTM A131 | A | I |
| Welded To | ASTM A131 | A | I |
| Backing Material | ASTM A131 | A | I |
| Other | | | |

| BASE METAL THICKNESS | As-Welded | With PWHT |
|---|---|---|
| CJP Groove Welds | 3/4–1-1/2 in | – |
| CJP Groove w/CVN | – | – |
| PJP Groove Welds | – | – |
| Fillet Welds | – | – |
| DIAMETER | – | – |

| JOINT DETAILS | |
|---|---|
| Groove Type | Single V Groove Butt Joint |
| Groove Angle | 35° included |
| Root Opening | 1/4 in |
| Root Face | – |
| Backgouging | None |
| Method | – |

JOINT DETAILS (Sketch)

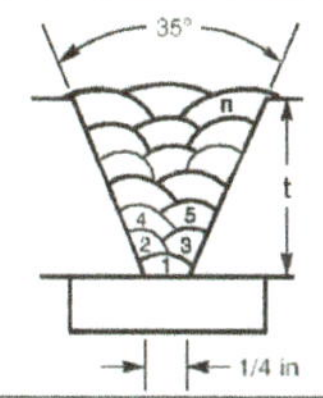


| POSTWELD HEAT TREATMENT | |
|---|---|
| Temperature | None |
| Time at Temperature | – |
| Other | – |

| PROCEDURE | | | | | | | | |
|---|---|---|---|---|---|---|---|---|
| Weld Layer(s) | All | | | | | | | |
| Weld Pass(es) | All | | | | | | | |
| Process | FCAW | | | | | | | |
| Type *(Semiautomatic, Mechanized, etc.)* | Semiauto | | | | | | | |
| Position | OH | | | | | | | |
| Vertical Progression | – | | | | | | | |
| Filler Metal (AWS Spec.) | A5.20 | | | | | | | |
| AWS Classification | E71T-1C | | | | | | | |
| Diameter | 0.045 in | | | | | | | |
| Manufacturer/Trade Name | – | | | | | | | |
| Shielding Gas (Composition) | 100% $CO_2$ | | | | | | | |
| Flow Rate | 45–55 cfh | | | | | | | |
| Nozzle Size | #4 | | | | | | | |
| Preheat Temperature | 60° min. | | | | | | | |
| Interpass Temperature | 60°–350° | | | | | | | |
| Electrical Characteristics | – | | | | | | | |
| Current Type & Polarity | DCEP | | | | | | | |
| Transfer Mode | – | | | | | | | |
| Power Source Type *(cc, cv, etc.)* | CV | | | | | | | |
| Amps | 180–220 | | | | | | | |
| Volts | 25–26 | | | | | | | |
| Wire Feed Speed | (Amps) | | | | | | | |
| Travel Speed | 8–12 ipm | | | | | | | |
| Maximum Heat Input | – | | | | | | | |
| Technique | – | | | | | | | |
| Stringer or Weave | Stringer | | | | | | | |
| Multi or Single Pass (per side) | Multipass | | | | | | | |
| Oscillation *(Mechanized, Automatic)* | – | | | | | | | |
| Number of Electrodes | 1 | | | | | | | |
| Contact Tube to Work Dist. | 1/2–1 in | | | | | | | |
| Peening | None | | | | | | | |
| Interpass Cleaning | Wire Brush | | | | | | | |

**Fig. 3.1** Example welding procedure specification (WPS)[1]

## 3.2 Procedure Qualification Record (PQR)

AWS B2.1 defines a PQR as a record of the welding variables used to produce an acceptable test weld and the results of the tests conducted on that weldment to qualify a welding procedure specification. ASME IX defines a PQR as a record that documents what occurred during the production of a procedure qualification test coupon and the

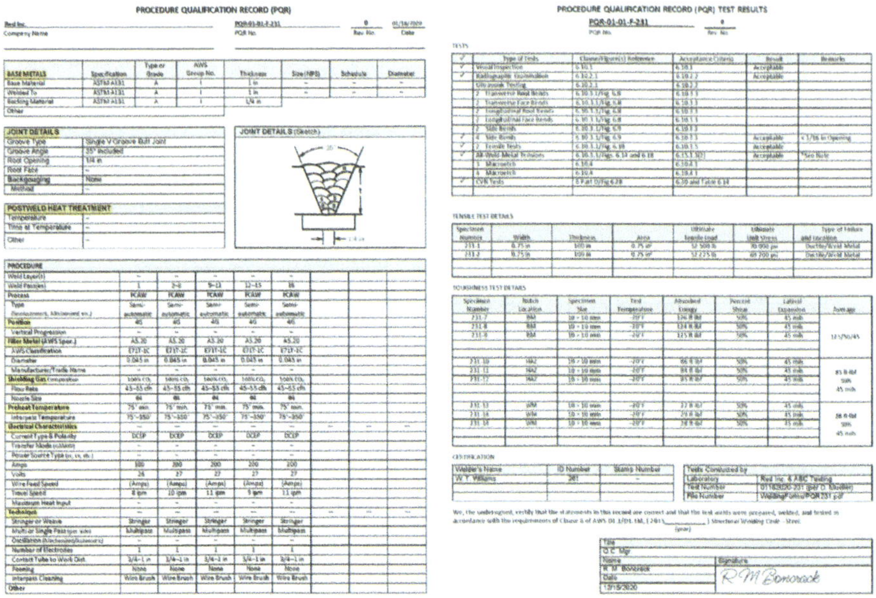


**Fig. 3.2** Example procedure qualification record (PQR)[1]

results of testing that coupon. The purpose of qualifying the procedure specification is to demonstrate that the joining process proposed for construction is capable of producing joints having the required mechanical properties for the intended application.

For convenience, the welding variables discussed in Chap. 5 are typically grouped into the categories similar to those for WPSs as stated above. An example of a Welding Procedure Specification is shown in Fig. 3.2 with variable groupings highlighted in yellow and test results shown on the right.

Similar to the WPS, any format may be used for the procedure qualification record provided all applicable information is recorded including a certifying statement acknowledging the validity of the data and certifying that the weldments were made and tested in accordance with the requirements. Examples of PQR formats are provided in AWS B2.1 Annex F Form F.3, and ASME IX Nonmandatory Appendix B Form QW-483.

## 3.3 Performance Qualification Record

The documentation of a performance qualification records what occurred during the production of a performance qualification test coupon by a person using one or more joining processes following an organization's welding procedure specification. AWS B2.1 labels this documentation as the Performance Qualification Test Record (PQTR) and provides an example form in Annex F. ASME IX labels this documentation as either the Welder Performance Qualification (WPQ) record, or Welder Operator Performance Qualification (WOPQ) record and provides example forms in Nonmandatory Appendix B.

For convenience, the welding variables discussed in Chap. 8 for performance qualification are typically grouped into the categories similar to those for WPSs as stated above. An example of a performance qualification record is shown in Fig. 3.3 with the columns for the welding variables, tested variable values, and qualified ranges highlighted in yellow along with the test results.

Similar to the WPS and PQR, any format for the performance qualification record may be used provided all applicable information is recorded including a certifying statement acknowledging the validity of the data and certifying that the weldments were made and tested in accordance with the requirements (ref. AWS B2.1 5.1.8 & AWS IX QW-301.4). Examples of performance qualification record formats are provided in AWS B2.1 Annex F Form F.1, and ASME IX Nonmandatory Appendix B Forms QW-484A and QW-484B.

## 3.4 Material Test Reports

The Material Test Report (MTR) is a quality assurance document utilized in the metals industry that certifies the material's chemical and physical properties meet or exceeds the standards required for a specific application to ensure safety. MTRs provide traceability and assurance to the end user about the origin and quality of the material and the process used to produce it.

MTRs are referred to by many synonymous names including a Mill Test Report or Metallurgical Test Report (MTR), Certified Mill Test Report or Certified Material Test Report (CMTR), Mill Test Certificate (MTC), Inspection Certificate, or Certificate of Test. Certification states that the manufacturer certifies they have complied with all the requirements of the material product specification (see Certificate of Compliance below).

The Material Test Report typically includes the following basic information.

Product Specification

Product specifications refer to the standards applied to a given material. A product specification may contain further sub-classifications or designations such as Grade, Type, and Class for steel base materials or various filler metal designations for welding consumables.

Examples of base material specifications include:

- American Society for Testing and Materials (ASTM) standards have a prefix including the letter "A"
- American Society of Mechanical Engineers (ASME) standards have a prefix including the letters "SA"
- European Norms (EN) standards have a prefix including the letters "EN".

PERFORMANCE QUALIFICATION TEST RECORD

| | | | | | |
|---|---|---|---|---|---|
| Name | Z. W. Elder | | Test Date | 12/12/2020 | Rev. |
| ID Number | 00-001-ZWE | | Record No. | WPQ-001 | 0 |
| Stamp No. | ZWE-1 | | Std. Test No. | ST-001 | 0 |
| Company | RED Inc. | | WPS No. | WPS-01-01-G-001 | 0 |
| Division | – | | Qualified To | AWS D1.1 | |

| BASE METALS | Specification | Type or Grade | AWS Group No. | Size (NPS) | Schedule | Thickness | Diameter |
|---|---|---|---|---|---|---|---|
| Base Material | ASTM A36 | UNS K02600 | I | – | – | 3/8 in | – |
| Welded To | ASTM A36 | UNS K02600 | I | – | – | 3/8 in | – |

| VARIABLES | Actual Values | RANGE QUALIFIED |
|---|---|---|
| Type of Weld Joint | Plate – Groove (Fig. 6.20) with Backing | Groove, Fillet, Plug, and Slot Welds (T-, Y-, K-Groove PJP only) |
| Base Metal | Group I to Group I | Any AWS D1.1 Qualified Base Metal |

| | Groove | Fillet | Groove | Fillet |
|---|---|---|---|---|
| Plate Thickness | 3/8 in | – | 1/8 in – 3/4 in | 1/8 in min. |
| Pipe/Tube Thickness | – | – | 1/8 in – 3/4 in | Unlimited |
| Pipe Diameter | – | – | 24 in min. | Unlimited |

| Welding Process | GMAW | GMAW |
|---|---|---|
| Type *(Manual, Semiautomatic, Mechanized, Automatic)* | Semiautomatic | Semiautomatic, Mechanized, Automatic |
| Backing | With | With (incl. Backgouging and Backwelding) |
| **Filler Metal (AWS Spec.)** | A5.18 | A5.xx |
| AWS Classification | ER70S-6 | All |
| F-Number | – | – |
| **Position** | 2G, 3G, and 4G | |
| Groove – Plate and Pipe □ 24 in | | All |
| Groove – Pipe < 24 in | | – |
| Fillet – Plate and Pipe □ 24 in | | All |
| Fillet – Pipe < 24 in | | All |
| Progression | Vertical Up | Vertical Up |
| GMAW Transfer Mode | Globular | Spray, Pulsed, Globular |
| Single or Multiple Electrodes | Single | Single |
| Gas/Flux Type | A5.32 SG-C | A5.xx Approved |

TEST RESULTS

| Type of Test | Acceptance Criteria | Results | Remarks |
|---|---|---|---|
| Visual Examination per 6.10.1 | 6.10.1 | Acceptable | – |
| Each Position: 1 Root Bend per 6.10.3.1 and Fig. 6.8 | 6.10.3.3 | Acceptable | – |
| Each Position: 1 Face Bend per 6.10.3.1 and Fig. 6.8 | 6.10.3.3 | Acceptable | 3G: Small (<1/16 in) Opening |

CERTIFICATION

| Test Conducted by | |
|---|---|
| Laboratory | ABC Testing |
| Test Number | 12122020-01 |
| File Number | Forms/12122020-01.pdf |

We, the undersigned, certify that the statements in this record are correct and that the test welds were prepared, welded, and tested in accordance with the requirements of Clause 6 of AWS D1.1/D1.1M (2015________) *Structural Welding Code—Steel.*
(year)

Manufacturer or Contractor: Red Inc. Authorized by: E. M. Ployee (Q.C. Mgr.)

Date: 12/12/2020

**Fig. 3.3** Example performance qualification test record (PQTR)[1]

Examples of filler metal specifications include:

- American Welding Society (AWS) standards have a prefix including the letter "A"
- American Society of Mechanical Engineers (ASME) standards have a prefix including the letters "SFA"
- International Organization for Standardization (ISO) standards have a prefix including the letters "ISO".

Material product specifications for the same material from differing SDOs are often but not always the identical. For example, the plane carbon steel specification for ASTM A36 and ASME SA36 are identical.

Material Heat Number

Material manufacturers can identify raw material in different ways, using lot, coil or other identifying numbers, but ultimately all MTRs will identify the material with a Heat Number. The heat number is used to maintain traceability of the material. When matching an MTR to its raw material all accompanying paperwork and in many cases markings on the raw material itself must match the heat number on the MTR.

Material Dimensions

The MTR identifies the applicable dimensions of the raw material. In the case of plate material this would be the thickness, round bar and pipe the diameter, or flat bar the thickness and width. This information must match the order requirements.

Mechanical Properties

There exists a large number of standardized tests that can be used to determine the various mechanical properties of materials. These include tensile testing, which is used to obtain the stress–strain curve for a material and thus properties such as Young Modulus, yield stress, tensile stress, and % elongation to failure. Charpy impact testing is a high strain-rate test which determines the amount of energy absorbed by a material during fracture. This absorbed energy is a measure of a given material's notch toughness and acts as a tool to study temperature-dependent ductile–brittle transition.

Chemical Analysis

Chemical Analysis, like mechanical properties, refers to a specific chemical makeup that certifies a material meets a particular product specification. The actual measured properties of the raw material are recorded on the MTR for the identified heat number. The values listed on the MTR must fall within the range or limits of the product specification for the raw material to be accepted for use.

Heat Treatment

Heat treatment is a group of thermal processes used to alter the physical, and sometimes chemical, properties of a material. The most common application is metallurgical. Heat treatment involves the use of heating or cooling, normally to elevated temperatures, to achieve the desired result such as hardening or softening of a material. Heat treatment techniques include annealing, case hardening, precipitation strengthening, tempering, carburizing, normalizing and quenching. The specific sequence of the heat treatment including temperatures, hold times, heating and/or cooling rates may be specified on an MTR.

CERTIFIED MATERIAL TEST REPORT

Page 1 of 1

| | | | |
|---|---|---|---|
| **Customer:** | Pressure Fabricators | | **Customer Part No:** SP-SA336F11CL32 |
| **Cust. P.O.:** | L529-1648 | **Item #:** 001 | **Tag:** |

| | |
|---|---|
| **Material Spec:** | ASME SECTION II PART A, SA336 F11 CL 3 & SA182 F11 CLASS 2, 2013 EDITION |
| **QA Desc** | NORMALIZED & TEMPERED |
| **Description:** | ROUGH MACHINED TO 2" TH X 36" WD X 36" LG |

**MILL DATA** **HEAT #** N9743 **MILL NAME:** AUTRY NATIONAL STEEL

| C | Mn | P | S | Si | Cr | Mo | Ni | V | Al | Cu | H | Ti | Sn | B | Nb | N |
|---|---|---|---|---|---|---|---|---|---|---|---|---|---|---|---|---|
| % | % | % | % | % | % | % | % | % | % | % | PPM | % | % | % | % | % |
| .14 | .44 | .009 | .011 | .53 | 1.08 | .47 | .23 | .004 | .026 | .12 | 1.0 | .002 | .007 | .0001 | .003 | .0063 |

**HEAT TREATMENT**

| Description | Temperature (F) | Soak Time (H.M) | Cooling | Load No | Other Info |
|---|---|---|---|---|---|
| NORMALIZE | 1650 | 12.00 | AIR COOL | 46588 | |
| TEMPER | 1150 | 6.15 | AIR COOL | 46620 | |

**MECHANICAL PROPERTIES - TENSILE**

| Yield (P.S.I.) | Ultimate (P.S.I.) | Elongation (%in4D) | Red of Area (%) | Offset (%) | Location | Direction | Comment |
|---|---|---|---|---|---|---|---|
| 46800 | 77500 | 32 | 67 | 0.2 | | | |

**ADDITIONAL QUALITY STATEMENTS**

FORGED IN 2 DIRECTIONS UNIFORMLY DUE TO UPSET FORGING
THE MATERIAL MEETS CLASS 1 & CLASS 2

***REVISED 12-17-16 - INCLUDE CLASS 1 & CLASS 2***
***REVISED 02-03-16 - DUAL CERTIFY TO SA182 F11 CLASS 2***

**TEST NUMBER:** 145697

**HARDNESS DETAILS:** 163 HBW

**SHIP WEIGHT:** 860 LBS

* NO WELD REPAIR WAS PERFORMED ON THESE PARTS.

* These forgings have not come into direct contact with mercury or any of its compounds, nor with any mercury containing device during manufacturing, testing, inspection or storage.

* All products supplied are in compliance with the quality requirements of the purchase order and specifications identified unless stated otherwise on this CMTR. All products outlined on this CMTR were produced in accordance with our quality manual.

**FORGED IN USA** Q.A. Assistant *Alison Suwatski* Alison Suwatski

Date: **May 18, 2018** QUALITY ASSURANCE

**Fig. 3.4** Example CMTR for SA-336 1.25Cr-0.5Mo-Si forging

Country of Origin and Manufacture

Frequently, the county of origin of the original material melt and subsequent processing of the ingot is provided.

Certified Mill Signature

The MTR may have a signature of a responsible employee of the foundry or mill producing the raw material, certifying that the information is accurate.

An example of a CMTR for a SA-336 1.25Cr-0.5Mo-Si forging is shown in Fig. 3.4, and a CMTR for ERNiCrMo-3 Inconel 625 filler metal is shown in Fig. 3.5.

It should be noted that sometimes a materials MTR may be missing. In these cases, a material may be properly identified by methods such as chemical analysis and mechanical testing. The results can then be reconciled with existing material specifications and added as supporting documentation.

**PATRIOT**
**Wire Corporation**

**PRODUCT CERTIFICATION**

| WORK ORDER | HEAT NUMBER |
|---|---|
| 1045422 | ZY224 |

SALES ORDER / RLS

SOLD TO:

CERT ID/REV

ISO 9001 Registered
AS:9100 Registered

| CUSTOMER P.O. | CUSTOMER PART | QUANTITY | COILS | LADING NO. | SHIPMENT DATE |
|---|---|---|---|---|---|
| | | 15340 LBS | 465 | S1874601 | 02/21/2018 |

DESCRIPTION
ALLOY 625, LEVEL WOUND, 0.045 DIN 300 Spool

CERTIFICATION REQUIREMENTS
UNS N06625
AMS 5837 G
AWS A5.14, 2011 & 2018, ERNiCrMo-3
BS EN10204:2004 3.1
ASME SECTION II, PART C, 2017 EDITION

**Chemical**

| AL | C | CO | CR | FE | MN | MO | NBTA | SI | TI |
|---|---|---|---|---|---|---|---|---|---|
| 0.10 | 0.01 | 0.02 | 22.23 | 0.26 | 0.03 | 8.70 | 3.693 | 0.07 | 0.19 |

| B | CA | CB/NB | CU | MG | N | NI |
|---|---|---|---|---|---|---|
| <0.0003 | <0.0008 | 3.69 | <0.01 | 0.0032 | 0.013 | 64.5 |

Chemistry Notes:
NICO - 64.52
P - <0.003
PB - <0.0001
S - <0.001
SN - <0.0002
TA - 0.0030
ZN - 0.0007
Melt - Imphy
Orig - France
TOE - <5

**Mechanical by Lot**

| Test | Units | High | Low | Average |
|---|---|---|---|---|
| Diameter | IN | 0.0434 | 0.043 | 0.0432 |

Visual and dimensional examination was satisfactory. Material, when shipped is free from contamination by mercury, radium, alpha source, and low melting point elements.

This is to certify that all required samplings, inspections and tests have been performed in accordance with the specification requirements. The test report represents the actual attributes of the material furnished and the values shown are correct and true. The material described by this certificate is in full compliance with all order and inspection requirements. We hereby certify that the above data are in accordance with the specification requirements.

Joe Qaqc - Quality Manager
Joe Qaqc

**Fig. 3.5** Example CMTR for ERNiCrMo-3 inconel 625 filler metal

## 3.5 Certificate of Compliance

A certificate of compliance, also referred to as a certificate of conformance, C of C, or CoC, is documentation that attests the material you receive meets specified requirements. All CoCs' state that the material provided meets the requirement requested in the purchase order or a specification, but they also generally include details like the date of the order, the specific items ordered including part numbers, and the origin and heat or lot number which can be used to help source the material should there be any issues. Sometimes, a statement of conformance is included on the Material Test Report (MTR).

## 3.6 Non-destructive Examination Reports

Non-Destructive Examination (NDE) are testing, and analysis techniques used to evaluate a qualification test coupon or mockup for the presence of defects and discontinuities without causing damage. NDT also known as non-destructive testing (NDT), non-destructive inspection (NDI) and non-destructive evaluation (NDE). The more common methods of NDE for procedure and performance qualification include Visual Inspection (VT), Liquid Penetrant Inspection (PT), Magnetic Particle Inspection (MT), Radiographic Inspection (RT), and Ultrasonic Inspection (UT).

A Non-Destructive Examination report will typically detail what was inspected, the applicable specifications, the inspection method, procedure, and calibration if applicable, the inspection results, and a signature certifying the results. An example of a liquid penetration examination (PT) report is shown in Fig. 3.6 along with an example of a radiography (RT) report in Fig. 3.7.

## 3.7 Destructive Examination Reports

Destructive Examination (DE), also known as Destructive Testing (DT) or Mechanical Testing, are testing and analysis techniques used to evaluate a qualification test coupon or mockup for physical or chemical properties and the presence of defects and discontinuities by physical destruction. The more common methods of DE for weld qualification include Macro Examination, Tension Tests, Guided-Bend Tests, Toughness Tests, and Hardness Tests.

A Destructive Examination report will typically detail what was inspected, the applicable specifications, the inspection method, procedure, and calibration if applicable, the inspection results, a signature certifying the results. An example of a report for tensile and bend testing is shown in Fig. 3.8.

## 3.8 Weld Technique Sheets

Weld Technique Sheets are supplemental information provided in addition to the Welding Procedure Specification in order to guide the welder/operator in completing the weld in production. While the WPS will address all required welding variables, it is common for the WPS to be somewhat generic in order to be utilized for multiple applications, allowing for a wide range of welding parameters, groove configuration, and welding technique. For

**ACME Inspection Services**

**LIQUID PENETRANT EXAMINATION REPORT**

Job # 15-053 PO # 68753 WO # N/A Date 8-25-2015 Page 1 of 1

MATERIAL: SA 178 Gr.A (ERNiCrMo-14) SFA5.14 N06686

PART SIZE: Mounted Samples 180° - 2, 3, 4, 5 & 6

**EXAMINATION PROCEDURE**

SPECIFICATIONS: ASME BPVC Section IX 2019 Edition

ACCEPT / REJECT CRITERIA: Customer Specification QW 453

| | | | |
|---|---|---|---|
| PENETRANT: MFG/DESIGNATION | Magnaflux 37 | BATCH NO. | 12E104 |
| EMULSIFIER: MFG/DESIGNATION | Magnaflux-10B | BATCH NO. | 07E080 |
| CLEANER: MFG/DESIGNATION | Magnaflux SKC-S | BATCH NO. | 12F03K |
| DEVELOPER: MFG/DESIGNATION | Magnaflux ZP-4B | BATCH NO. | 12E032 |

| | | | |
|---|---|---|---|
| PENETRANT DWELL TIME | 30 Minutes | PART TEMPERATURE | 65°F |
| EMULSIFIER DWELL TIME | N/A | TYPE | 1 |
| DRYING TIME | 10 Minutes | METHOD | D |
| DEVELOPING TIME | 10 Minutes | SENSITIVITY | 4 |
| POST CLEAN | Solvent Wipe | LIGHTING EQUIPMENT | Black light (ACME 03282) |

**EXAMINATION RESULTS**

| IDENTIFICATION | ACCEPT | REJECT | QUANTITY | REMARKS |
|---|---|---|---|---|
| Section 2 | √ | | 1 | No Indications Noted |
| Section 3 | √ | | 1 | No Indications Noted |
| Section 4 | √ | | 1 | No Indications Noted |
| Section 5 | √ | | 1 | No Indications Noted |
| Section 6 | √ | | 1 | No Indications Noted |
| | | | | |
| | | | | |
| | | | | |
| | | | | |
| | | | | |
| | | | | |
| | | | | |
| | | | | |
| | | | | |

INSPECTION PERFORMED BY: Siegfried Fischbacher Level II P.T.

Roy Horn Level III P.T.

CLIENT APPROVAL: Albert Einstein

**Fig. 3.6** Example NDE report for liquid penetrant examination

specific applications, a WPS may be supplemented with a Weld Technique Sheet which prescribes very specific instructions. An example of a technique sheet for an orbital GTAW pipe weld on a high-temperature alloy is shown in Fig. 3.9. Specific details limiting the joint dimensions and prescribing the welding sequence and parameters for each layers are provided.

Other types of technique sheets may provide specific instructions on welding machine settings. An example of a technique sheet for the second layer of a machine GTAW temper bead weld is shown in Fig. 3.10 where specific machine settings are provided for a Liburdi Dimetrics—Gold Track II power supply. In the case of a temper bead weld,

**ACME Inspection Servicces**

**RADIOGRAPHIC INSPECTION REPORT**

Job # 356892 PO # 0102494 WO # N/A Date 6/7/2021 Page 1 of 1

CLIENT: WE Co. Welding
LOCATION: 1248 Arthur E Adams Dr.
Colum bus, OH 43220

PART NO.: 1.750" x 0.390"
PART NAME: Pipe Weld Sample
MATERIAL: N/A
THICKNESS: 0.390"
TYPE WELD: Full Penetration Butt Weld

**RADIOGRAPHIC INSPECTION TECHNIQUE**

| | | | |
|---|---|---|---|
| Specification(s) | ACME 20.1 Rev. 28 | Acceptance Criteria | ASME B31.3 Severs 2018 Ed. |
| Isotope | N/A | Film Mfg/Type | AGFA D4 D5 |
| Curies | N/A | Film Size | 4.5" X 17" |
| KV | 300 | Sensitivity | Wire # 7 (0.013") |
| MA | 5 | Image Quality Indicator | ASTM 1B (Source Side) |
| Time | 54 Seconds | Shim(s) | N/A |
| SFD | 36" | Develop: Temp: | 83°F Time: Auto |
| SOD | 34.126" | Screen(s) | Pb Front & Back 0.010" |
| OFD | 1.874" | Films Per Cassette | 2 |
| Source Size (Focal) | 0.157" | Weld Reinforcement Thickness | 0.062" |
| Geometric Unsharpness | 0.008" | Welding Process | N/A |

**INTERPRETATION**

| PART I.D./ FILM VIEW | Accept | Reject | Crack | Slag Inclusion | Tungsten Inclusion | Porosity | Undercut | Incomplete Fusion | Incomplete Penetration | Root Concavity | Root Convexity | DENSITY Image Quality Indicator IQI | DENSITY Area of Interest | REMARKS |
|---|---|---|---|---|---|---|---|---|---|---|---|---|---|---|
| 5G | | | | | | | | | | | | | | |
| Sample A | | | | | | | | | | | | | | |
| 0 | √ | | | | | | | | | | | 2.4 | 2.6 | Weld Profile Noted |
| 60 | √ | | | | | | | | | | | 2.5 | 2.7 | |
| 120 | √ | | | | | | | | | | | 2.4 | 2.6 | |
| | | | | | | | | | | | | | | |
| 5G | | | | | | | | | | | | | | |
| Sample B | | | | | | | | | | | | | | |
| 0 | | X | | | | | | X | | | | 2.3 | 2.6 | Add. Film for Verification |
| 60 | √ | | | | | | | | | | | 2.4 | 2.5 | |
| 120 | √ | | | | | | | | | | | 2.2 | 2.3 | |

**SETUP**

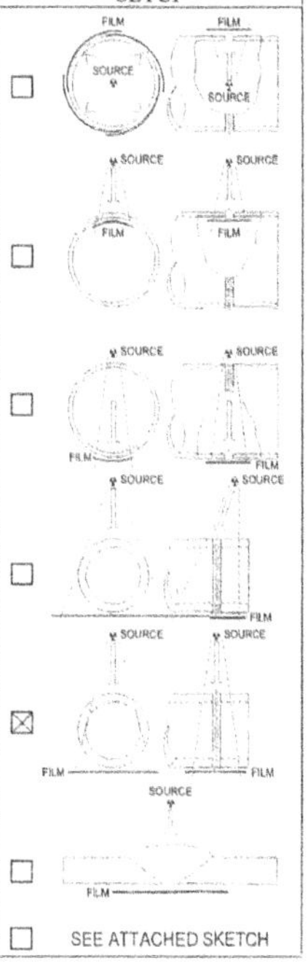


RADIOGRAPHER(S): Kaygee Graff Level II R.T. DATE: 6/7/2021
INTERPRETER: Kaygee Graff Level II R.T. DATE: 6/7/2021

**Fig. 3.7** Example NDE report for radiography

each layer may have its own technique sheet outlining the welding parameters to be used with that layer.

## 3.9 Weld Bead Logs

Weld Bead Logs are commonly created during the welding of a qualification coupon in order to record the actual welding parameters. Information such as preheat and inter-pass temperatures, welding current, arc voltage, travel speed, and wire feed speed if applicable.

ACME Testing Inc.

**CERTIFIED TEST REPORT**

| | | | |
|---|---|---|---|
| **CLIENT:** | WD Welding Co. | **CLIENT #:** | 7332 |
| **ADDRESS:** | 1375 E. Buena Vista Dr. | **PROJECT #:** | 22579 |
| **CITY, STATE, ZIP:** | Orlando, Fl. 43830 | **PURCHASE ORDER #** | WSI0098567 |
| **ATTENTION:** | Walt Disney | **EVALUATION DATE:** | August 16, 2019 |

Material Description: 1.5" thick ASTM A871 Grade 60 to ASTM Grade 60 using GTAW with ERNiCrMo-3.
Specification: PQR 2528

TENSILE RESULTS –ASME Section IX 2019

| Sample ID | Thickness (Inch) | Width (Red. Inch) | Area (sq. inch) | Yield Load (lbf) | Yield Strength (psi) .2% Offset | Ultimate Tensile Load (lbf) | Ultimate Tensile Strength (psi) | % Elongation (2-inch) | Fracture Location |
|---|---|---|---|---|---|---|---|---|---|
| Transverse Tensile A | .500 | 0.750 | 0.3750 | 5,090 | 14,157 | 27,816 | 74,178 | 18 | Base Material |
| Transverse Tensile B | .500 | 0.750 | 0.3750 | 5,309 | 17,817 | 30,105 | 80,280 | 19 | Base Material |
| Transverse Tensile C | .500 | 0.750 | 0.3750 | 5,117 | 16,980 | 29,340 | 76,132 | 18 | Base Material |

| Sample ID | Result |
|---|---|
| Side Bend A | Accepted |
| Side Bend B | Accepted |
| Side Bend C | Accepted |
| Side Bend D | Accepted |

Summary of Testing: Reported results pertain only to samples submitted for testing. The samples were tested for information purposes only.

Tested By: Donald Duck, Laboratory Technician

Reviewed By: Will Smith, Laboratory Manager

**Fig. 3.8** Example DT report for tensile and bend testing

This information can be as detailed as for every individual weld bead, every weld layer, or less detailed such as for the root and hot passes, fill passes, and cap passes. Additional information such as all non-essential variables for the Procedure Qualification Record should also be recorded. An example of a Weld Bead Log is shown in Fig. 3.11.

The Weld Bead Log might also contain a Weld Bead Map as shown in Fig. 3.12. The Weld Bead Map may depict the joint configuration and dimensions as well as the placement of all weld beads and their sequence within the weld joint.

## 3.10 Post-weld Heat Treatment Reports

Some materials require the qualification coupon have a Post-Weld Heat Treatment (PWHT) performed. The actual requirements may vary by material and are typically described in the applicable fabrication or construction standards. Whomever performs

**WELDING TECHNIQUE SHEET**

**WeCo Welding LLC**
**1248 Arthur E Adams Dr.**
**Columbus OH 43221**

Page 1 of 1

Project Petro-Americas Lubricants
Welding Procedure Specification No. WPS-45-KHR35CT-T-001 rev.0 Date: 06/09/2020
Welding Process(es) Machine GTAW

Scope This Welding Technique Sheet shall be used to restrict or define variable ranges for the referenced Welding Procedure Specification (WPS) being utilized for the referenced project. Any deviation from these limitations and ranges shall require written authorization from Welding Engineering.

JOINT DETAILS (QW-402)

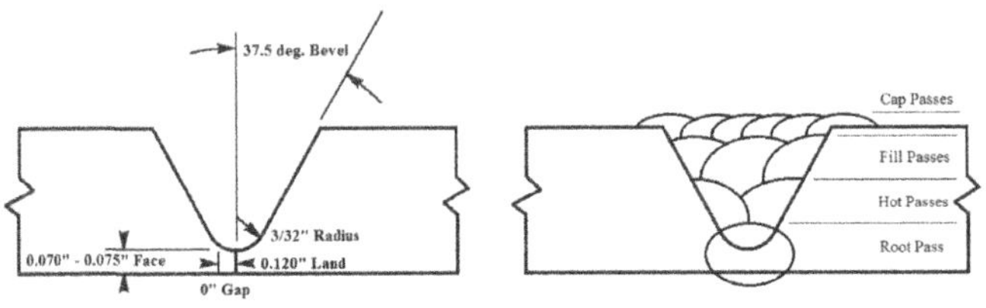


| | | Filler Metal | | Welding Parameters | | | |
|---|---|---|---|---|---|---|---|
| Weld Layer(s) | Process | Class | Diameter | Amperage Range⑤ | Voltage Range | Travel Speed Range | Wire Feed Speed Range |
| Root Pass | GTAW Machine | ERNiCrMo-3 | 0.035" | 135 / 70 | 9.6 | 1.0 IPM | 35 IPM |
| Hot Pass(es) | GTAW Machine | ERNiCrMo-3 | 0.035" | 145 / 85 | 9.6 | 1.5 IPM | 25 - 45 IPM |
| Fill Pass(es) | GTAW Machine | ERNiCrMo-3 | 0.035" | 145 / 90 | 9.9 - 10.2 | 1.7 - 1.8 IPM | 25 - 45 IPM |
| Cap Pass(es) | GTAW Machine | ERNiCrMo-3 | 0.035" | 140 - 150 / 90 | 10.0 - 10.3 | 1.8 - 1.9 IPM | 25 - 45 IPM |

**NOTES**
① (QW-402.2) Backing material shall not be used.
② (QW-404.22) Consumable inserts shall not be used.
③ (QW-406.4) The maximum interpass temperature shall be 350°F maximum.
④ (QW-408.5) Backing gas shall be used.
⑤ (QW-409.3) Pulsed current shall be used.

**Fig. 3.9** Weld technique sheet for an orbital GTAW pipe weld

the PWHT should supply a report which contains material details for traceability as well as the PWHT parameters such as heating and cooling rates, soak temperature and time, as well as a graph of the actual PWHT temperature history usually recorded on a strip chart and measured by thermocouples. An example of a PWHT report is shown in Fig. 3.13.

**Technique Data Sheet for Second Temperbead Layer**

(Applicable only to Layer 2)

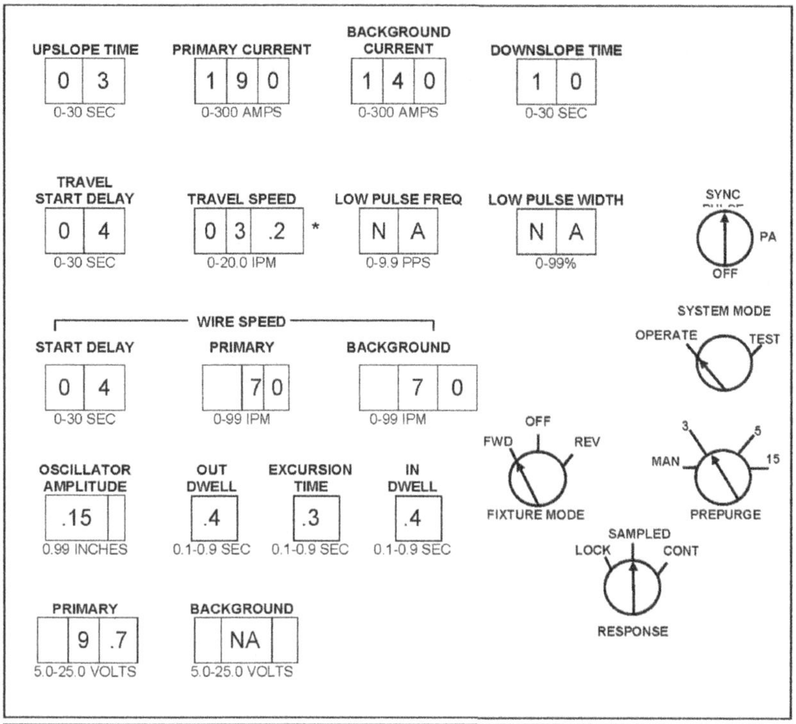


**Average Amperage**=(Primary Amps*Pulse Width) + Background Amps*(1-Pulse Width) = 168.6 Amps
**Average Voltage**=(Primary Volts*Pulse Width) + Background Volts*(1-Pulse Width) = 8.9Volts
**Average Heat Input Rate (KJ per inch) = (Average Amperage*Average Voltage*60) *10$^{-3}$ ÷ Travel Speed = 28.1 KJ/In.**
**Power ratio: 71 KW/In$^2$**
**Layer 2 heat input rate must be within the range of 26.9 KJ/In minimum through 32.9 KJ/In maximum. Parameters may be adjusted as long as the heat input stays within this range.**

NOTES: *** For 28.1 KJ/In travel speed dial setting is to be calculated for 3.2 IPM @ the tungsten**
**Tungsten lead angle is 10 degrees. No torch tilt. Place tungsten between 1/10" and 1/16" from fusion line on this layer to develop proper bead shape. Initiate beads on the lowest elevation of the area to be covered by the pad.**

**Fig. 3.10** Weld technique sheet for layer 2 of a machine GTAW weld

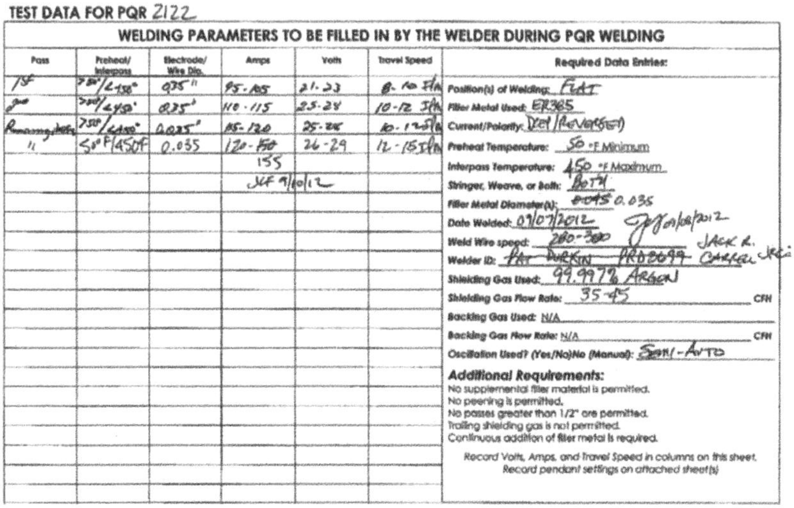

TEST DATA FOR PQR 2122

WELDING PARAMETERS TO BE FILLED IN BY THE WELDER DURING PQR WELDING

| Pass | Preheat/ Interpass | Electrode/ Wire Dia. | Amps | Volts | Travel Speed |
|---|---|---|---|---|---|
| 1st | 750°/<450° | .035" | 95-105 | 21-23 | 8-10 I/M |
| 2nd | 750°/<450° | .035" | 110-115 | 25-28 | 10-12 I/M |
| Remaining [illegible] | 750°/<450° | 0.035" | 115-120 | 25-28 | 10-12 I/M |
| " | 50°F/450F | 0.035 | 120-150 155 | 26-29 | 12-15 I/M |
| | | | JLF 9/10/12 | | |

Required Data Entries:

Position(s) of Welding: FLAT
Filler Metal Used: ER365
Current/Polarity: DC+ (Reverse)
Preheat Temperature: 50 °F Minimum
Interpass Temperature: 450 °F Maximum
Stringer, Weave, or Both: BOTH
Filler Metal Diameter(s): ~~.045~~ 0.035
Date Welded: 09/07/2012
Weld Wire speed: 280-300
Welder ID: ~~PAT DURKIN PR03699~~ JACK R. CARREL
Shielding Gas Used: 99.997% ARGON
Shielding Gas Flow Rate: 35-45 CFH
Backing Gas Used: N/A
Backing Gas Flow Rate: N/A CFH
Oscillation Used? (Yes/No)No (Manual): SEMI-AUTO

**Additional Requirements:**
No supplemental filler material is permitted.
No peening is permitted.
No passes greater than 1/2" are permitted.
Trailing shielding gas is not permitted.
Continuous addition of filler metal is required.

*Record Volts, Amps, and Travel Speed in columns on this sheet.*
*Record pendant settings on attached sheet(s)*

**Fig. 3.11** Example weld bead log

**Fig. 3.12** Example weld bead map

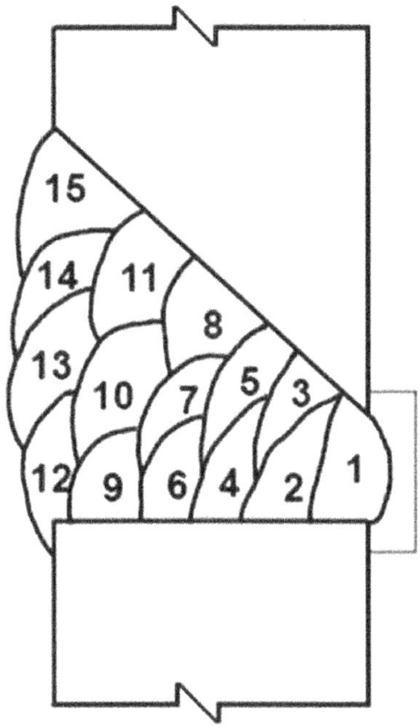

## ACME HEAT TREATING

### POST WELD HEAT TREATMENT REPORT

**Ref.** Job#355429 **Date** May 26, 2020 **Page** 1 **of** 1

**Heat Treatment Details**

Material: 2" OD x 0.136" Wall SA-213 T11 with ERNiCrMo-10 CRO applied at 0.055" thick.
Specifications: Customer Supplied.
Procedure: Post weld heat treat was performed at 1250°F +/- 25°F with a heating and a cooling rate of 300°F/hr maximum above 600°F for 30-45 minutes.

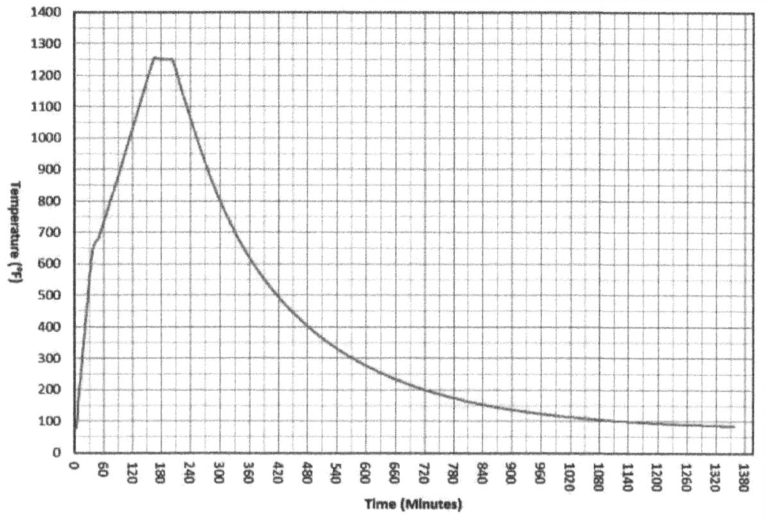


Prepared by: Audry Hepburn
H.T.Technician

Approved by: Cary Grant
H.TSupervisor

**Fig. 3.13** Example PWHT report

## 3.11 Summary

Most manufacturers or organizations quality management program or system typically has requirements for record keeping including identification, traceability, and records retention. Additional requirements may be imposed by invoked standards such as ASME BPVC IX where QW-103.2 requires each organization shall maintain a record of the results obtained in welding procedure and welder and welding operator performance qualifications. Another example is AWS D1.1 where 6.3.3 states record of the test results shall be kept by the manufacturer or contractor and shall be made available to those authorized to examine them. All the documents discussed in this chapter are considered key records related to welding qualification.

### Note

1. 2020, American Welding Society, Structural Welding Code—Steel, AWS D1.1/D1.1 M:2020, Annex J.

# Welding Procedure Development and Qualification Process

# 4

The process of developing a welding procedure and then qualifying it can be both complicated and time consuming. The general workflow of the process is shown in Fig. 4.1. Depending on the complexity of the welding procedure and required qualification testing, the process may be as simple as modifying an existing welding procedure or simply using a Standard or Pre-Qualified welding procedure to as complex as developing a welding procedure from scratch. Details for each step of the process are discussed separately below.

## 4.1 Requirements Review

Prior to initiating the procedure qualification process, all the requirements should be reviewed in preparation for the PQR. In general, the requirements can be classified as Application specific Requirements, Qualification Requirements, or Other Considerations. These requirements summarized in Table 4.1, will be discussed in detail in the following sections.

### 4.1.1 Production Weld Type

The weld type is a classification based on its basic shape and joint design. The weld type that will be utilized in production will determine the configuration and testing rules for the qualification test weldment.

© The Author(s), under exclusive license to Springer Nature Switzerland AG 2025

D. Barborak, *Arc Welding Qualification Standards*, Synthesis Lectures on Welding Engineering, https://doi.org/10.1007/978-3-031-64646-1_4

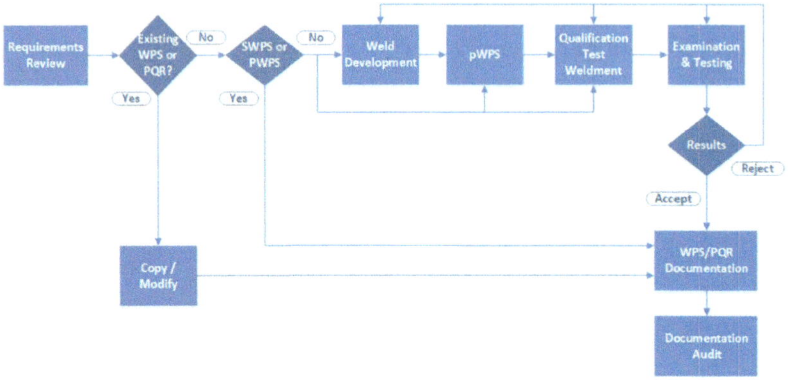


**Fig. 4.1** Workflow of the welding procedure development and qualification process

**Table 4.1** Procedure qualification requirements

| Application requirements | Qualification requirements | Other considerations |
|---|---|---|
| • Production weld type<br>• Production welding position<br>• Production weldment thickness<br>• Welding process<br>• Materials | • Governing qualification standards<br>• Customer/contractual/ jurisdictional requirements<br>• Welding variables<br>• Testing and examination requirements<br>• Test weldment type | • Resource constraints<br>• Weld induced distortion<br>• Welding sequence |

Below are the basic weld type classifications for arc welding, along with their required test weldment configurations.

- Groove Welds—groove joint configuration
- Fillet Welds—fillet or groove joint configuration
- Plug/Slot Welds—groove joint configuration
- Spot Welds—groove joint configuration
- Seam Welds—fillet or groove joint configuration
- Flange Welds—fillet or groove joint configuration
- Surfacing Welds—bead-on-plate configuration.

For a complete description of the various weld type classifications, refer to AWS A3.0 Standard Welding Terms and Definitions.

### 4.1.2 Production Weldment Thickness

The thickness of the production weldment will determine the required base metal thickness and weld deposit thickness of the qualification test weldment. Determining the thickness requirements of the production weldment is usually straight forward, but sometimes complex part geometries can make determination of the governing thickness more challenging. For the case of multiple components, members, or parts that have differing thicknesses, all thicknesses must be considered. Examples include butt joints and T-joints between dissimilar thickness workpieces. Certain components such as flange assemblies or valve bodies may have complex geometries where the governing thickness may be difficult to determine depending on the location of the weld.

The qualification thickness ranges are provided in AWS B2.1 Table 4.3 and ASME IX Table QW-451.1. It should be noted that components with different thicknesses may require multiple PQRs to qualify a WPS in order to cover all thicknesses. Further discussion of the thickness requirements for groove and fillet weld procedure qualification are provided in Chap. 5, for corrosion resistant overlay procedure qualification in Chap. 6, for hardfacing overlay procedure qualification in Chap. 7, and for welder/operator performance qualification in Chap. 8.

### 4.1.3 Production Welding Position

Welding position is defined as the relationship between the weld pool, joint, joint members, and welding heat source during welding. Depending on the welding process and consumables utilized, welding position can significantly affect the resulting weld quality and mechanical properties of a weldment. There are essentially four different fundamental welding orientations, namely flat, horizontal, vertical, and overhead positions. Vertical position welding can be carried out as vertical upward progression or vertical downward progression. Specific designations are given to each position for fillet welds (e.g. 1F, 2F, 3F, 4F, 5F, 6F) and for groove welds (e.g. 1G, 1GR, 2G, 3G, 4G, 5G, 6G). Some positions such as 5G and 6G are a combination of orientations namely flat, vertical, and overhead. A description of the various welding positions are provided in AWS B2.1 Annex A1 and ASME IX QW-120.

While a procedure or performance qualification standard may allow qualification in any position to qualify for all positions, considerations should be given to the welding positions that will be required in production. Some welding processes have all-position capabilities using unique techniques and specially designed welding consumables, while others may be used in only one or two positions. The welding process, parameters, technique, and consumables should be suitable for use in the positions permitted by the procedure or performance qualification.

Further discussion of the positional requirements for groove and fillet weld procedure qualification are provided in Chap. 5, for corrosion resistant overlay procedure qualification in Chap. 6, for hardfacing overlay procedure qualification in Chap. 7, and for welder/operator performance qualification in Chap. 8.

### 4.1.4 Welding Process

Several welding processes may be applicable for a particular application. The selection of an appropriate joining process for a given application involves numerous considerations including the following[1]:

- Availability and fitness for service
- Skill requirements
- Weldability of the base metal alloy with respect to type and thickness
- Availability of suitable welding consumables
- Weld joint design
- Heat input requirements
- Demands of the welding position
- Economics of the process, including capital expenditures
- Materials and labor
- Number of components being fabricated
- Applicable code requirements, and
- Safety concerns.

Selecting the best process is often a compromise of several factors. The requirements for procedure and welder/operator qualification are generally addressed by joining process category, namely Arc Welding, Resistance Welding, Solid-State Welding, High Energy Beam Welding, Brazing, Soldering, or other miscellaneous processes such as Electroslag Welding. While specific welding processes within a given category may have many similar welding variables to consider, all processes also have unique variables that must be considered and addressed separately during qualification of each process.

### 4.1.5 Materials

Most welding standards assign alphanumeric designations to base materials based on their weldability because of the large number of alloy specifications that exist in order to reduce the number of procedure and performance qualifications required. These designations are assigned to base metals dependent on characteristics such as chemical composition, mechanical properties, metallurgical compatibility, and weldability. Both AWS B2.1 and

**Table 4.2** Comparison of base metal groupings (list is not comprehensive)

| Material type/nominal composition | AWS B2.1 | | | ASME IX | | |
|---|---|---|---|---|---|---|
| | M-number | # Groups | # Spec/Cl./Gr. | P-number | # Groups | # Spec/Cl./Gr. |
| Carbon steel | 1 | 4 | 835 | 1 | 4 | 197 |
| 0.5Cr–0.5Mo | 3 | 4 | 92 | 3 | 3 | 65 |
| 1.25Cr–0.5Mo | 4 | 3 | 50 | 4 | 2 | 39 |
| 2.25Cr–1Mo | 5A | 1 | 35 | 5A | 1 | 27 |
| 5-9Cr–0.5Mo | 5B | 1 | 33 | 5B | 2 | 34 |
| 5-9Cr–0.5Mo–V | 5C | 5 | 38 | 5C | 5 | 38 |
| 9Cr–1Mo–V (P91) | 15E | 1 | 15 | 15E | 1 | 20 |
| Austenitic stainless | 8 | 4 | 634 | 8 | 4 | 476 |
| Duplex stainless | 10H | 1 | 97 | 10H | 1 | 43 |
| Ni alloys | 41–49 | N/A | 426 | 41–49 | N/A | 331 |
| Al alloys | 21–27 | N/A | 241 | 21–26 | N/A | 69 |
| Ti alloys | 51–53 | N/A | 236 | 51–53 | N/A | 88 |

ASME IX base these designations on material specifications from ASTM, ARS, API, AS, NZS, CSA, AMS, ISO, NACE, and MSS. AWS B2.1 assigns a Material or "M" Number which are listed in Normative Annex C. Similarly, ASME IX assigns a "P" number which are listed in QW-420. When toughness requirements are specified, the ferreous materials are further subdivided into group numbers. A comparison of the alphanumeric designations, number of groups, and number of different specifications, classifications, and grades for select material types are given in Table 4.2.

Similar to base materials, most welding standards group welding filler metals based on their usability characteristics to reduce the number of procedure and performance qualifications required. These groupings do not imply that filler metals within a group may be indiscriminately substituted for the filler metal was used in the qualification test coupon without consideration of the compatibility of the base metals and filler metals metallurgical compatibility, PWHT requirements, design, service requirements, and mechanical properties. Both AWS B2.1 and ASME IX similarly assign a Filler or "F" Number which are listed in AWS B2.1 Normative Annex B and ASME IX QW-430. A summary of the F number designations and number of specifications for select filler metal electrodes and rods are given in Table 4.3.

Finally, many welding standards group the deposited weld metal chemistry for ferrous weld metals. Both AWS B2.1 and ASME IX similarly assign an "A" Number which are listed in AWS B2.1 Normative Annex B and ASME IX QW-442. Note the A number does not refer to the filler wire or electrode chemical composition but instead refers

**Table 4.3** Summary of filler metal groupings

| Material type | AWS B2.1 and ASME IX | |
|---|---|---|
| | F-Number(s) | # Specifications |
| Steel and steel alloys | 1–6 | 16 |
| Aluminum and aluminum alloys | 21–26 | 2 |
| Copper and copper alloys | 31–37 | 4 |
| Nickel and nickel alloys | 41–46 | 4 |
| Titanium and titanium alloys | 51–56 | 1 |
| Zirconium and zirconium alloys | 61 | 1 |
| Hardfacing weld metal overlay | 71–72 | 2 |

to the deposited weld metal chemical composition. The filler metal and deposited weld metal will have different chemical compositions due to dilution with the base material during welding. The A number is usually provided by the filler metal manufacturer who perform chemical analysis on the deposited weld metal. Alternatively, the A number can be determined during procedure qualification.

While the M or P numbers, and F numbers are essential variables for both procedure and performance qualification, the A numbers are only essential variables for procedure qualification and not applicable to performance qualification. Further discussion of the base metal, filler metal, and weld deposit requirements for groove and fillet weld procedure qualification are provided in Sect. 4.5, for corrosion resistant overlay procedure qualification in Sect. 4.6, for hardfacing overlay procedure qualification in Sect. 4.7, and for welder/operator performance qualification in Sect. 4.8.

### 4.1.6 Shielding Gas and/or Flux Composition

During arc welding nitrogen, oxygen and hydrogen can react with the molten metal, causing defects and a degradation in mechanical properties that weaken the weld. The primary function of a shielding gas is to protect the molten weld metal from atmospheric contamination while it solidifies. The shielding gases may be generated either by the decomposition of fluxing materials or by their direct introduction into the arc stream and the area surrounding the arc plasma. In addition to its shielding function, each flux, gas or blend has unique physical properties that can have a major effect on welding speed, penetration, mechanical properties, weld appearance and shape, fume generation, weld color, and arc stability.

The primary gases used for arc welding are argon, helium, hydrogen, nitrogen, oxygen and carbon dioxide. Shielding gases are used in either a pure form or in blends of varying components. The selection of a gas or gas mixture can become quite complex due to the many combinations available. Arc welding fluxes are compositionally more complex and typically contain silicates and metal oxides and may contain flux binders, ferro-alloys and metal powders. The composition of the flux or gas can and should be tailored to meet the process, material, and application requirements. Selection of a flux shielding gas or blend must be based on a knowledge of what is available, their applications, and the overall effect they have on the welding process.

A change is shielding gas or flux composition is usually considered an "essential variable" in most standards that govern welding procedure and performance qualification. Further discussion of the shielding gas and flux compositional requirements for groove and fillet weld procedure qualification are provided in Chap. 5, for corrosion resistant overlay procedure qualification in Chap. 6, for hardfacing overlay procedure qualification in Chap. 7, and for welder/operator performance qualification in Chap. 8.

### 4.1.7 Governing Qualification Standards

The rules for qualification are typically dictated by the governing standards which varies by industry and/or application. These governing standards will either contain all the qualification requirements itself or refer to a separate stand-alone qualification standard, sometimes known as support codes or standards.

Examples of stand-alone support standards that contain all the qualification requirements include:

- AWS B2.1 "Specification for Welding Procedure and Performance Qualification"
- ASME IX "Welding, Brazing, and Fusing Qualifications".

Codes, standards, specifications, or contract documents that invoke a stand-alone qualification standard such as AWS B2.1 or ASME Sec. IX are considered referencing documents. These reference documents, sometimes referred to as construction codes, typically contain requirements such as materials used and their mechanical properties such as tensile strength and toughness, joint design, preheating, postweld heat treatment (PWHT), acceptance criteria for weld quality, and related examinations. While the requirements within referencing documents take precedence over those of a stand-alone qualification standard, they may impose additional qualifications requirements above and beyond those contained in the referenced stand-alone qualification standard.

Examples of standards that refer to a separate stand-alone qualification standard:

- AWS D14.3 "Specification for Welding Earthmoving, Construction, Agricultural, and Ground-Based Material Handling Equipment" requires qualification to AWS B2.1
- ASME BPVC Section VIII Division 1 "Rules for Construction of Pressure Vessels" requires qualification to ASME IX.

Some standards such as AWS D1.1 "Structural Welding Code—Steel" has its own provisions for qualification but will also allow qualifications to other standards with an engineer's approval. In another example, AWS B2.1 4.1.3 states, "Tests previously conducted by an Employer to meet other codes, specifications, or earlier editions of this specification, may be used by the Employer to support a WPS in accordance with this specification. The Procedure Qualification Record(s) (PQR) must address all essential variables and if required, supplementary essential variables that are applicable to the welding process(es) used and the test results must meet all requirements of this specification. If all the requirements are not met, another qualification test may be conducted following the qualification variables of the original PQR. A PQR supplement shall be prepared to document the additional tests results."

*Qualification to Multiple Standards*

When performing a single welding procedure or performance qualification to multiple standards, considerations should be given to:

- Are the test requirements and acceptance criteria the same?
- Are the essential variables and their qualification ranges similar or compatible?

In most cases, the more stringent requirements typically govern and sometimes duplicate testing to separate test requirements need to be performed. To qualify a welding procedure to more than one qualification standard using a single coupon, the test requirements of both standards must be met when the coupon is tested. The essential variables and supplementary essential variables when required for each standard must be documented on the PQR(s). The WPS must meet the essential/supplementary essential variable rules for each standard and the nonessential variables (when specified) for each standard must also be addressed on the WPS.

Because the rules of each standard are made by different committees that serve different industries whose regulatory requirements and technical concerns vary, the rules have been developed independently resulting in significant differences between the standards.[2]

The following are some comparisons of the differing qualification test requirements between different standards[2]:

- Procedures qualified under AWS D1.1, D1.6, and B2.1 as well as ASME IX do not require a nick-break test in addition to the required bend and tension tests for a groove procedure qualification while API 1104 does require the nick-break test. Therefore, procedures qualified under the AWS or ASME standards cannot be used for API 1104 pipeline construction.
- In addition, AWS D1.1, D1.6, B2.1, and ASME IX require the weld reinforcement be removed from a tensile specimen while API 1104 does not. Therefore, a tensile specimen with weld reinforcement, but having an internal defect, could pass an API

1104 tension test, whereas it could have failed a test conducted to one of the other standards.

- For complete joint penetration (CJP) groove welds, AWS D1.1 requires radiographic testing (RT). A welding procedure qualified under AWS D1.6, B2.1 or ASME IX would not meet this requirement because RT is not required under these standards, therefore their procedures could not be used.
- For bend tests, AWS D1.1 allows flaws upto a 0.125 in. [3.2 mm] long on the surface with a total cumulative length of 0.375 in. [9.5 mm] of acceptable flaws while ASME IX and AWS B2.1 permit an unlimited number of 0.125 in. [3.2 mm] long flaws on the surface. Therefore, a bend test that passes ASME IX or AWS B2.1 may not pass the requirements of AWS D1.1 because of the limit on total cumulative length.
- For tensile tests, ASME IX and AWS B2.1 permit a specimen to pass provided the tensile strength is not more than 5% below the minimum tensile strength required for the weaker of the base metals being joined. AWS D1.1 nor AWS D1.2 have this provision while API 1104 has a similar but not identical rule.

The following is an example of a procedure qualification to both ISO and ASME BVPC requirements:

- Visual Examination: ISO would govern (more comprehensive)
- Liquid Penetrant Examination: ISO would govern (not required by ASME)
- Radiographic Examination: ISO would govern (not required by ASME)
- Tensile Testing: ISO would govern (ASME size requirements are more liberal)
- Bend Testing: ASME and ISO are same for $t <= 0.375$ in. [9.5 mm] and $t > 0.472$ in. [12 mm]
- Macroetch Testing: ISO would govern (not required by ASME)
- Hardness Testing: May require separate testing to both ASME and ISO requirements (depends on the criteria invoked)
- Charpy Impact Testing: Requires testing to both ASME and ISO requirements (requirements are not compatible).

### 4.1.8 Customer/Contractual/Jurisdictional Requirements

In addition to the governing standards, additional requirements may be invoked by the customer, contractually, or by a jurisdiction. These requirements may be invoked specifically, or by reference to other standards.

Examples of customer specific or contractual requirements include:

- The requirement that welding consumables utilized for a PQR and for the actual work be purchased from one of their approved vendors.

- The requirement that a WPS be qualified with the same Heat or Lot of filler metal that will be used in production or repair.
- The requirement that the weld deposit chemistry or FN meets a stated specification at a given depth. This requirement may be imposed for the PQR and/or for production welding.
- The requirement that Low Hydrogen Electrodes (<4 mg/100 mL) be utilized.
- The requirement that Hardness Testing be performed for the PQR and/or for production welding. The test requirements and acceptance criteria may be from a separate SDO such as API, NACE, or ISO, etc.

Examples of jurisdictional requirements include:

- The requirement that a welding procedure be registered and/or reviewed by the jurisdictional authorized inspector (AI) the work is being conducted in.
- The requirement that welder qualifications be witnessed by a jurisdictional authorized inspector (AI) the work is being conducted in.

### 4.1.9 Welding Variables

Welding variables are elements, features, or factors that can vary or change and can affect the mechanical integrity or quality of a completed weldment. A variable, whose change may affect the mechanical properties (other than toughness) of a weldment is considered an Essential Variable. A variable, whose change may affect the toughness properties of a weldment, heat-affected zone, or base material is considered a Supplementary Essential Variable. Supplementary essential variables become additional essential variables in situations where procedure qualifications require toughness testing. When procedure qualification does not require the addition of toughness testing, supplementary essential variables are not applicable. A variable, whose change does not affect the mechanical properties of a joint is considered a Non-Essential Variable. Non-Essential Variables shall be addressed in the Welding Procedure Specification, as required. In most qualification standards, welding variables are generally grouped by welding process as each process may have its own unique variables while have other common variables with similar processes. When preparing for a welding procedure or performance qualification, all the required Essential and Supplementary-Essential variables if required should be considered.

*Essential Variables*

Essential variables are those variables in which a change, as described in the specific variables, is considered to affect the mechanical properties of the weldment. If there is a change in the essential variable the procedure must be requalified. Examples include a change in Welding Process, Base Material Classification (P or M number), or Filler Metal Classification (F number).

*Supplementary Essential Variables*
Supplementary essential variables are required for metals for which other Sections or Codes specify notch-toughness testing and are in addition to essential variables for each process. This means that when ASME Section VIII (which requires qualification to ASME IX) also requires notch toughness testing on a material, the supplementary essential variables become essential variables for that WPS. A change in a supplementary essential variables requires re-qualification of the procedure. Examples include a change in Base Material Group Number, or Filler Metal Classification within a Specification.

*Non-essential Variables*
Non-essential variables are those in which a change, as described in the specific variables, may be made in the WPS without re-qualification. Examples include a change in Groove Design, or Filler Metal Diameter.

*Welding Variable Categories*
Welding variables are generally categorized along with the typical information and data as follows:

- Joint Design
  - Joint configuration and dimensions
  - Backing.
- Base Metal(s)
  - Material number and subgroup
  - Thickness range qualified
  - Diameter (pipe and tube).
- Filler Metal(s)
  - Specification and classification
  - F- and A-Number or nominal chemical composition
  - Weld metal deposit thickness
  - Filler metal size or diameter
  - Flux classification
  - Supplemental filler metal
  - Consumable insert and type.
- Position(s)
  - Welding position(s)
  - Progression for vertical welding.
- Preheat and Interpass Temperature
  - Minimum preheat temperature

  - Maximum interpass temperature
  - Preheat maintenance.
- Post Weld Heat Treatment
  - PWHT temperature and time.
- Shielding Gas
  - Torch shielding gas composition and flow rate range
  - Root shielding gas composition and flow rate range.
- Electrical Characteristics
  - Current (or wire feed speed), current type, polarity, pulse characteristics (if applicable)
  - Voltage range
  - Type and diameter of tungsten electrode (GTAW)
  - Transfer mode.
- Other Variables/Technique
  - single or multi electrodes and spacing
  - Single-pass or multi-pass
  - Contact tube to work distance
  - Cleaning
  - Peening
  - Stringer/weave bead (manual welding) or oscillation (mechanized or automatic welding)
  - Travel speed range.

A comparison of the number of welding variables required for a SMAW groove and corrosion resistant overlay (CRO) procedure qualification are shown in Table 4.4 for AWS B2.1 and ASME IX.

Further discussion of the welding variables for groove and fillet weld procedure qualification are provided in Sect. 5.1, for corrosion resistant overlay procedure qualification in Sect. 6.1, for hardfacing overlay procedure qualification in Sect. 7.1, and for welder/operator performance qualification in Sect. 8.3.

**Table 4.4** SMAW welding variables for a groove PQR and a CRO PQR

| | AWS B2.1 | ASME Sec. IX |
|---|---|---|
| *# Groove variables* | | |
| Essential | 18 | 9 |
| Supplementary essential | 8 | 10 |
| Non-essential | 23 | 17 |
| *# CRO variables* | | |
| Essential | 23 | 10 |
| Non-essential | 23 | 5 |

### 4.1.10 Test Weldment Type

In addition to a standard test weldment for procedure and performance qualification, AWS B2.1 allows several types of special test weldments (ref. AWS B2.1 4.3.7). ASME IX only allows special test weldments for fillet welds (ref. ASME IX QW-181.1.1) and tube-to-tubesheet welds (ref. ASME IX QW-193.1). The testing requirements and acceptance criteria of special test weldments are typically defined in the governing standards or reference documents. Each type of test weldment are discussed below.

*Standard Test Weldments*

These are test weldments in which the coupon is configured to perform the basic required destructive testing. Examples of standard qualification test weldments such as fillet, groove, hardfacing, and corrosion resistant overlay are discussed in Sect. 4.1.11 below.

It should be noted that both AWS B2.1 and ASME IX allows other standard qualification test weldment configurations such as pipe or welded box tubing for groove weld qualifications. Refer to the applicable qualification standards for the coupon configuration and test requirements.

*Simulated Service Test Weldments*

These are special test weldments in which qualification requires tests simulating service conditions. Tests types include fracture toughness, static or cyclic loading, hydrostatic loading, or corrosion testing.

An example of a simulated service test is the use of a Cruciform Fatigue Test to simulate cyclic fatigue loading of fillet welds (see Fig. 4.2). The size, quality, and geometry of the fillet weld, specifically at its weld toes, will determine the fatigue life of the weldment.

Another example of a simulated service test is a Ballistic Test in accordance with NATO standard STANAG 4569 where a 0.30 in. [7.62 mm] round projectile is fired at the test weldment from 32.8 yards [30 m] at 760 yards/sec. [695 m/s] (Fig. 4.3).

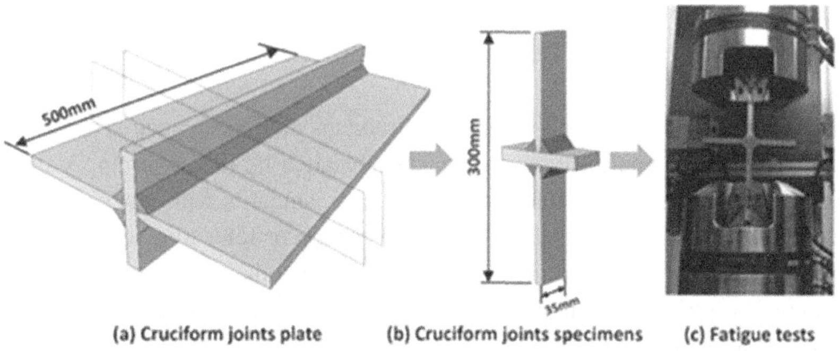


**Fig. 4.2** Cruciform fatigue testing of fillet welds[3]

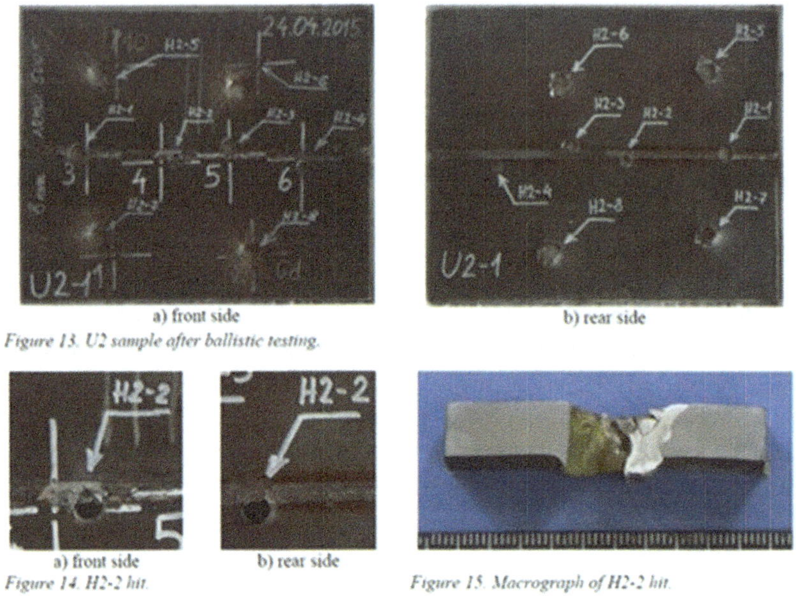


**Fig. 4.3** Ballistic test of an ARMOX 500 T steel welded joint[4]

*Prototype Test Weldments*

These are special test weldments in which a prototype of the actual weldment is subjected to field tests in which it is loaded and demonstrated to perform the function for which it was designed.

An example of a prototype test weldment is a hydrostatic "burst" test in which a welded prototype of a pressure vessel is subjected to elevated pressures. Depending on the test criteria, the prototype may be subjected to failure. An example of a small pressure vessel subjected to Hydrostatic testing is shown in Fig. 4.4.

### 4.1.11 Examination and Testing Requirements

The subsequent examination and testing requirements and acceptance criteria for procedure and performance qualification are typically dictated by the governing standards. As with the qualification requirements, these governing standards will either state the examination and testing requirements and acceptance criteria or refer to other standards which identify such requirements. The most common test methods include tension testing used to determine the ultimate strength of the test weldment, bend testing used to determine the degree of soundness and ductility of a test weldment, and toughness testing used to determine the toughness of a weldment. The minimum size of the test weldment coupon is

**Fig. 4.4** Hydrostatic "Burst" test of a small pressure vessel

highly dependent on the type of qualification being performed (e.g. groove, fillet, overlay, etc.), the test coupon product form (e.g. plate or pipe), and the number of test specimens required (e.g. tension, bend, toughness, etc.). References for the required dimensions of bend and tension test specimens can be found in Table 4.5. The required dimensions for other tests such a charpy impact test specimens for toughness testing can either be found in the referencing document or code of construction.

**Table 4.5** Test specimen dimensions

| Test | AWS B2.1 | ASME IX |
|---|---|---|
| Transverse face and root bend specimens | Figure A.2A | Figure QW-462.3(a) |
| Longitudinal face and root bend specimens | Figure A.2B | Figure QW-462.3(b) |
| Transverse side bend specimens | Figure A.2C | Figure QW-462.2 |
| Rectangular reduced section tension specimens | Figure A.3A | Figures QW-462.1(a) and QW-462.1(b) |
| Round reduced section tension specimens | Figure A.3B | Figure QW-462.1(d) |
| Alternate reduced section tension specimens for pipe | Figure A.3C | Figure QW-462.1(c) |
| Alternate full section tension specimens for pipe<= 2 in. [51 mm] | Figure A.3D | Figure QW-462.1(e) |

References for the required test specimen locations can be found in Table 4.6. The required locations for other less common qualification test types such as fillet weld configurations and qualification on box tubing can also be found in AWS B2.1 and ASME IX.

An example of a more complex groove weld qualification is shown in Fig. 4.5. Here, the qualification requires 2 transverse tensile tests, 2 all weld metal hot tensile tests, 4 transverse bend tests, 6 Charpy impact tests, 4 macros and 1 hardness traverse. The total required test coupon dimension is 12 × 24 in. [30.5 × 61 cm].

When preparing for a welding procedure qualification, all of the required examination and testing requirements should be reviewed in order to properly size the qualification

**Table 4.6** Test specimen locations

| Qualification test type | AWS B2.1 | ASME XI |
|---|---|---|
| *Procedure qualification* | | |
| Groove weld in pipe with transverse face and root bends | Figure 4.2 | Figure QW-463.1(d) |
| Groove weld in pipe with transverse side bends | Figure 4.2 note 2 | Figure QW-463.1(e) |
| Groove weld in plate with transverse face and root bends | Figure 4.5 | Figure QW-463.1(a) |
| Groove weld in plate with transverse side bends | Figure 4.5 | Figure QW-463.1(b) |
| Groove weld in plate with longitudinal face and root bends | Figure 4.4 | Figure QW-463.1(c) |
| Corrosion resistant overlay on plate | Figure 4.8 | Figure QW-462.5(d) |
| Harfacing overlay on plate | Figure 4.9 | Figure QW-462.5(e) |
| *Performance qualification* | | |
| Groove weld in pipe with transverse face and root bends | Figures 5.3 and 5.4 | Figure QW-463.2(d) |
| Groove weld in pipe with transverse side bends | Figures 5.3 and 5.4 note 1 | Figure QW-463.2(e) |
| Groove weld in plate with transverse face and root bends | Figure 5.6 | Figure QW-463.2(a) |
| Groove weld in plate with transverse side bends | Figure 5.6 note 2 | Figure QW-463.2(b) |
| Groove weld in plate with longitudinal face and root bends | Figure 5.7 | Figure QW-463.2(c) |
| Corrosion resistant overlay on plate | Figure 5.12 | Figure QW-462.5(d) |
| Harfacing overlay on plate | Figure 5.13 | Figure QW-462.5(e) |

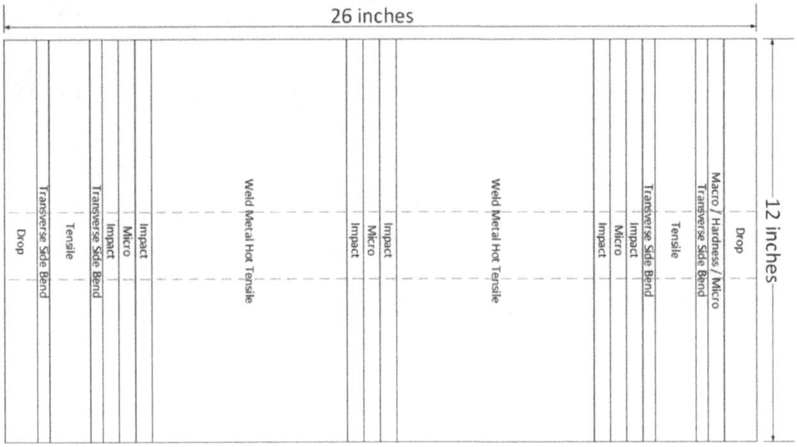


**Fig. 4.5** Example of a complex groove procedure qualification coupon requiring multiple tests

test weldment coupon. Further discussion of the examination and testing requirements and acceptance criteria for groove and fillet weld procedure qualification are provided in Sect. 5.2, for corrosion resistant overlay procedure qualification in Sect. 6.2, for hard-facing overlay procedure qualification in Sect. 7.2, and for welder/operator performance qualification in Chap. 8.

*Test Coupon Material Loss*

In addition to the required test specimen size discussed above, consideration should also be given to material loss. For test weldments produced in plate, most standards dictate that the ends be discarded. For example, AWS B2.1 recommends 0.5 in. [12.7 mm] and ASME IX recommends 1 in. [25.4 mm]. The use of run-on/run-off tabs at the ends of the joint as discussed in the next section below can be utilized to minimize the amount discarded at the ends.

In addition to discarding the ends, material loss from the test specimen cutting process should be considered. This loss, called Kerf, varies by the cutting process with typical values for the most common cutting processes as follows:

- Plasma: 0.150 in. [3.81 mm]
- Band Saw: 0.093–0.125 in. [2.36–3.18 mm]
- Abrasive Disk: 0.093–0.156 in. [2.36–3.96 mm]
- Oxy-Fuel: 0.045 in. [1.14 mm]
- Waterjet: 0.035 in. [0.89 mm]
- Laser: 0.025 in. [0.64 mm].

**Fig. 4.6** Plate groove joint with run-on/run-off tabs

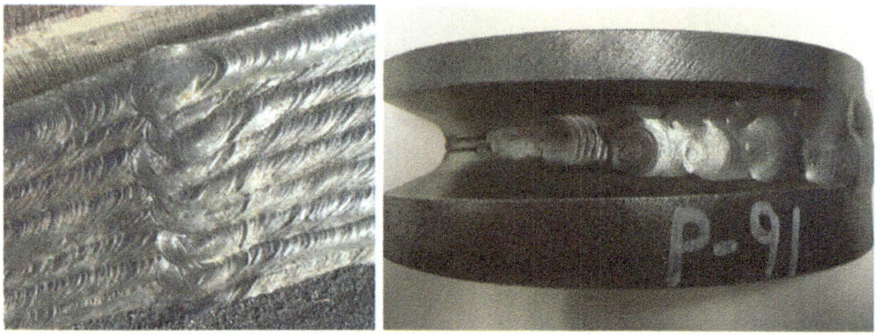


**Fig. 4.7** Overlapping starts/stops (left), staggered starts/stops (right)

*Weld Starts and Stops*
For multi-pass welds, consideration should also be given to weld starts (initiation) and weld stops (termination). Depending on the amount of material available and the number of destructive test specimens to be extracted from the test weldment, the use of run-on/run-off tabs may be required. Run-on/run-off tabs may be as simple as using an extended backing plate as shown in Fig. 4.6 (left), or more elaborate as shown in Fig. 4.6 (right).

For test weldments produced in pipe, overlapping weld starts and stops as shown in Fig. 4.7 (left) will increase risks for discontinuities and should be avoided. For groove welds in pipe, the weld starts and stops should be staggered as shown in Fig. 4.7 (right).

### 4.1.12 Resource Constraints

Consideration should also be given to the resources available to complete the procedure qualification and implement in production. Available resources may include:

- Equipment: Specific welding process equipment and/or fixturing
- Material: Base Material and Filler Metal
- Skilled Labor: Qualified welders for a given process with up-to-date continuity.

### 4.1.13 Weld Induced Distortion

Most test weldments will distort in one or more dimensions. The six main forms of weld induced distortion for a butt-weld in a plate include transverse shrinkage, longitudinal shrinkage, rotational distortion, angular distortion, bending distortion, and buckling distortion. The magnitude of distortion is primarily determined by the weld heat input, number of passes, groove configuration, and plate thickness. If the magnitude of the distortion is severe, it may be difficult to properly extract test specimens for destructive examination. Techniques to minimize the amount of distortion include pre-setting the weld joint, balanced welding progression, and the use of restraints.

By pre-setting the weldment, the parts are pre-set by a pre-determined amount and left free to distort during welding in order to achieve overall alignment and dimensional control. Examples of pre-set configurations include pre-setting a fillet joint to prevent angular distortion, pre-setting a butt joint to prevent angular distortion, and pre-setting a butt joint with a tapered gap to prevent joint closure as the weld progresses. These techniques may require some experimentation because it is difficult to predict the amount of pre-setting required. The main advantages compared to using restraints are that there is no expensive fixturing needed and there will be lower residual stress in the structure.

Because of the in-exact science of applying pre-set, the use of restraint is more common for procedure qualification test weldments. The basic principle is that the test weldment is placed in position and held under restraint to minimize any movement due to distortion during welding. Welding with restraint will generate additional residual stresses in the weld which may cause cracking. When welding highly hardenable materials, the use of elevated preheat and optimized welding sequence will help minimize the risk for cracking. When removing the restraint after welding, a relatively small amount of distortion can occur due to residual stresses. This distortion can be mitigated by either applying a small amount of pre-set or stress relieving heat treatment before removing the restraint.

Restraint is relatively simple to apply using clamping fixtures or jigs, and/or welded strongbacks to hold the parts during welding. An example of restraining a test weldment by the use of both welded strongbacks and clamps are shown in Fig. 4.8.

Balanced welding progression on a multi-pass 2-sided fillet or butt weld is also an effective means of controlling angular distortion. The welding sequence is staggered by welding on alternating sides of the joint to ensure that angular distortion is continually being countered and not allowed to accumulate during welding.

**Fig. 4.8** Test weldment utilizing clamps and strongbacks for restraint

## 4.2 Use of Existing WPSs and/or PQRs

One option may be to modify an existing Welding Procedure Specification or write a new WPS based on an existing Procedure Qualification Record. If the current application and its requirements do NOT require changes to any essential variables then you can modify the existing WPS, making needed changes to the non-essential variables. In some cases, you may want to write a new WPS based on an existing PQR. The PQR will then support multiple WPSs as discussed below in Sect. 4.5.1.

Depending on the application requirements, an additional qualification test coupon may need to be welded and tested in order to address supplementary requirements. As an example, you can supplement an existing qualified groove welding procedure without impact testing by running another test coupon under similar conditions and then performing impact testing. The existing WPS can then be modified to be applicable to both non-impact tested and impact tested applications. Another example would be supplementing an existing welding procedure qualified without hardness testing by running another test coupon under similar conditions and then performing hardness testing.

If the current application requirements DOES require changes to any essential variables then you can use the existing WPS parameters as a starting point, but must qualify a new welding procedure by running a new procedure qualification.

## 4.3 Use of a Prequalified WPS (PWPS)

Several AWS standards committees established provisions for the use of Prequalified Welding Procedure Specifications or PWPSs based on years of experience with more common welding processes, base metals, filler metals, and joint designs that have historically been qualified successfully many times and proven to work. The use of a PWPS alleviates the time and expense of performing additional qualification testing of these

**Table 4.7** PWPS limitations for several AWS standards

| AWS standard | Joint types | Welding processes | Materials (Spec., type, grade, and class) |
|---|---|---|---|
| D1.1 structural welding code—steel | 72 | SMAW, SAW, GMAW (except GMAW-S), FCAW | ~137 structural steels |
| D1.3 structural welding code—sheet steel | 9 | SMAW, SAW, GMAW (including GMAW-S), FCAW | ~130 sheet steels |
| D1.4 structural welding code—reinforcing steel | 1 (fillet) | SMAW, SAW, GMAW (except GMAW-S), FCAW | ~7 reinforcing steels |
| D1.5 bridge welding code | 58 | SMAW, SAW, GMAW, FCAW | ~8 steels |
| D1.6 structural welding code—stainless steel | 70 | SMAW, SAW, GMAW, FCAW, GTAW | ~329 stainless steels |

combinations. The implementation and use of a PWPS is free, only requiring the contractor to generate its own written WPS that meets all the governing code's requirements and limitations, no procedure qualification testing is required. Typical limitations include the combination of welding process, base and filler materials, joint details, preheat and interpass temperatures, welding parameter ranges, and PWHT. A summary of several AWS standards that allow PWPSs and some of their limitations are shown in Table 4.7.

It should be noted that the ASME Boiler and Pressure Vessel codes and Piping codes do not allow the use of Prequalified WPSs.

## 4.4 Use of a Standard WPS (SWPS)

Another option for developing a WPS is the use of a Standard Welding Procedure Specification or SWPS. The American Welding Society publishes SWPSs which are developed by the Welding Procedures Committee of the Welding Research Council (WRC). It is the intent of the AWS B2 Committee to have SWPSs only for commonly welded materials and common manual and semiautomatic welding processes. There are currently 80 SWPSs available for purchase from AWS. A summary of the SWPSs available for purchase is shown in Table 4.8.

The data to support a SWPS are derived from Procedure Qualification Records (PQRs) generated by the Welding Procedures Committee of the WRC and PQRs received from industry and government agencies. For all committee generated PQRs, the welding, testing, and evaluation are performed under the direct supervision of the WRC Welding

**Table 4.8** Summary of available SWPSs

| Application | # SWPSs | Materials | Thicknesses | Processes |
|---|---|---|---|---|
| Plate | 14 | Carbon steel<br>Stainless steel | 0.125-inch thru<br>1.50-inch | SMAW<br>GTAW<br>FCAW and FCAW-G |
| Pipe | 35 | Carbon steel<br>Stainless steel<br>Cr–Mo steels | 0.125-inch thru<br>1.50-inch | SMAW<br>GMAW-S<br>GTAW<br>FCAW-G |
| Sheet metal | 13 | Carbon steel<br>Galvanized steel<br>Stainless steel<br>Aluminum | 0.050-inch thru<br>0.141-inch | SMAW<br>GMAW-S<br>GTAW<br>FCAW |
| Naval (plate and pipe) | 18 | Carbon steel<br>Stainless steel | 0.125-inch thru<br>1.50-inch | SMAW<br>GTAW<br>FCAW-G |

Procedures Committee. An SWPS and its supporting PQRs must meet the rules for qualification of AWS B2.1, meet the rules of the major codes which govern the intended applications, and be approved by the AWS B2 Committee on Procedure and Performance Qualification.

As an example, part of the front matter of AWS SWPS B2.1-1-027 is shown in Fig. 4.9. This FCAW SWPS for plate and structural applications is supported by 26 PQRs.

The range of conditions and variables listed for an SWPS be more restrictive than permitted by application of the full range of conditions and variables allowed by the B2.1 document or by other American National Standards. The purpose of this conservatism is to restrict the WPS to a high probability of successful application by all users. While the minimum number of supporting PQRs required is two, but is usually in the range of two to fifty, the specific minimum number being a Committee decision for each SWPS.

The AWS Structural Codes D1.1, D1.2, D1.3, and D1.6, as well as the National Board Inspection Code (NBIC) for the repair of ASME code stamped items, permit the use of these SWPSs "as is". ASME IX, however, permits the use of 33 accepted SWPSs and requires the organization using them to conduct a demonstration test prior to use. The demonstration test is required by ASME IX because many of the codes that reference ASME IX, such as the ASME B31 piping codes and the API 620 and 650 standards do not require a third party involvement for production welding, whereas the National Board does require the involvement of an authorized inspection agency for the repair of ASME code-stamped items. A list of the SWPSs permitted for use in ASME IX are listed in its Mandatory Appendix E.

**<u>Standard Welding Procedure Specification (SWPS)</u>**

**Self-Shielded Flux Cored Arc Welding of Carbon Steel (M-1 or P-1, Groups 1 and 2), 1/8 [3 mm] through 1/2 inch [13 mm] Thick, E71T-11, As-Welded Condition, Primarily Plate and Structural Applications**

**Welding Research Council—Supporting PQR Numbers: 007014, 007015, 007016, 200015, 200293, 200295, 200296, 200431, 200432, 200433, 200434, 200600, 200601, 200602, 200603, 200608, 200609, 200610, 200731, 200732, 200733, 200787, 200790, 200791, 200794, 200796**

**Requirements for Application of SWPSs**

**Scope.** The data to support this Standard Welding Procedure Specification (SWPS) have been derived from the above listed Procedure Qualification Records (PQRs) which were reviewed and validated under the auspices of the Welding Research Council. This SWPS is not valid using conditions and variables outside the ranges listed. The American Welding Society considers that this SWPS presents information for producing an acceptable weld using the conditions and variables listed. The user needs a significant knowledge of welding and accepts full responsibility for the performance of the weld and for providing the engineering capability, qualified personnel, and proper equipment to implement this SWPS.

**Application.** This SWPS is to be used only as permitted by AWS B2.1, *Specification for Procedure and Performance Qualification*, and the applicable fabrication document(s) [such as code, specification, or contract document(s)]. The fabrication document(s) should specify the engineering requirements such as design, need for heat treatment, fabricating tolerances, quality control, and examination and tests applicable to the end product.

**Fig. 4.9** Example front matter of AWS SWPS B2.1-1-027

## 4.5 General Requirements

In addition to the considerations and requirements discussed above, the following are other general requirements to consider when combining multiple welding procedure specifications, multiple procedure qualification records, or qualifying multiple welding processes in a single test weldment. Additional discussion is also provided for simultaneous procedure and performance qualification, and simultaneous procedure qualification by multiple organizations, and organizational oversight.

### 4.5.1 A Single PQR Supporting Multiple WPSs

Many standards allow the use of one PQR to qualify more then one WPS. An example of this is shown in Fig. 4.10 where a single SMAW groove PQR is used to qualify three WPSs in accordance with AWS B2.1. Because the PQR was qualified with impact testing, WPS #1 can be written to be used for any non-impact tested application which allows for use with any P-No.1 base material with any Group number, a minimum thickness of 1/16 in. [1.5 mm], and use of any F-No.2 A-No.1 electrode. Because the PQR was qualified with impact testing, WPS #2 can be written to be used for any impact tested application which limits the base material to P-No.1 G-No.2 only, a minimum thickness of 5/8 in. [16 mm], and E6013 electrodes only.

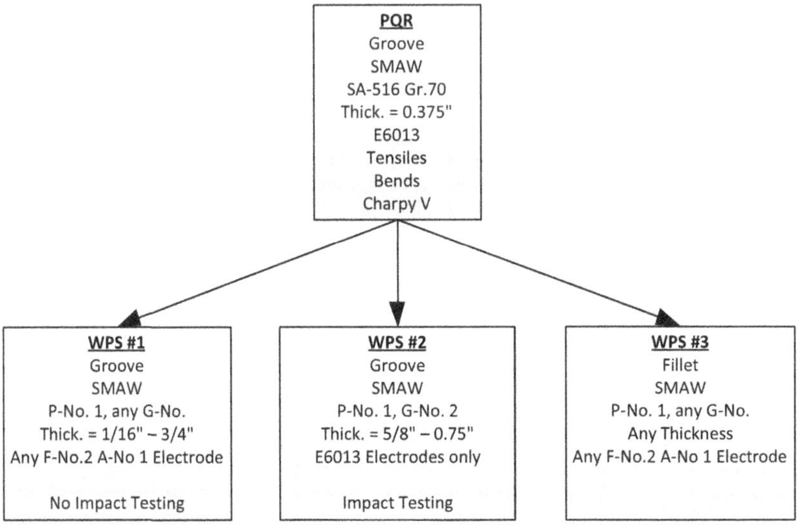


**Fig. 4.10** Single PQR supporting multiple WPSs

## 4.5.2 Multiple PQRs Supporting a Single WPS

Many standards also allow the use of multiple PQRs to qualify one WPS. An example is shown in Fig. 4.11 where a groove GTAW PWR and a groove SMAW PQR are used to qualify a WPS that requires the root and hot passes be performed by GTAW open-root, followed by fill and cap passes performed by SMAW.

**Fig. 4.11** Multiple PQRs supporting a single WPS

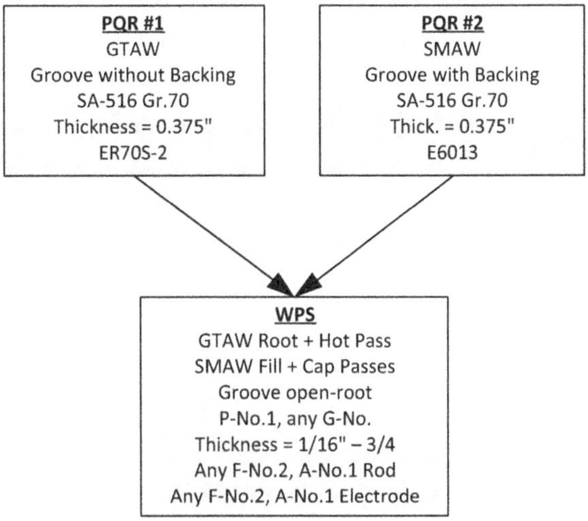

### 4.5.3 WPSs Supporting Multiple Welding Processes

Many qualification standards allow the mixed use of multiple welding processes. A multiple process WPS requires addressing all essential variables for both processes. Qualification of a multiple process WPS can be accomplished with multiple PQRs qualified with each process individually, or a single PQR where the processes are used in combination to complete the test weldment. AWS B2.1 4.5.4 further states "the thickness ranges permitted for use in the WPS shall apply separately to each welding process and filler metal classification. The weld deposit thickness for each welding process and each filler metal classification used in the qualification test shall be recorded on the WPS.

A good example of a multiple process WPS is a "Tig/Stick" (GTAW/SMAW) pipe welding procedure whose intention is to support welding a pipe groove weld with a GTAW root and hot pass, while performing the fill and cap passes with SMAW. This procedure can be qualified by a single PQR test coupon by welding the root and hot passes by GTAW, followed by welding the fill and cap passes with SMAW. The PQR coupon is then subsequently tested by tensile and bend testing and a single PQR is then written. An alternative way to qualify such a procedure would be to qualify separate PQRs for each process to be utilized, such as the example given in Fig. 4.11.

### 4.5.4 Simultaneous Procedure and Performance Qualification

Both AWS B2.1 and ASME IX allow simultaneous welding procedure qualification along with a welder or welding operator performance qualification on the same test weldment (ref. AWS B2.1 5.1.12 and ASME IX QG-106.2(e), QW-301.2). A welder or welding operator who completes an acceptable welding procedure qualification test are qualified within the limits of the performance qualification variables tested (see Chap. 8).

### 4.5.5 Simultaneous Procedure Qualifications by Multiple Organizations

ASME IX allows multiple organizations to collectively perform procedure qualification of one or more welding procedures simultaneously. Each participating organization must have a representative present to provide oversight of the simultaneous procedure qualification. The provisions and limitations of the simultaneous procedure qualifications are provided in ASME IX QG-106.4. AWS B2.1 has no provisions for simultaneous performance qualifications by multiple organizations.

### 4.5.6 Procedure Qualification Oversight and Organizational Responsibilities

Both AWS B2.1 and ASME IX require the procedure qualification process, including the welding of the qualification test weldment, be under full oversight by the qualifying organization (ref. AWS B2.1 4.3.20 and ASME IX QG-106). Personnel employed by the qualifying organization are responsible for supervision, control, evaluation, and acceptance of qualification testing. This oversight ensures compliance with the applicable qualification requirements. Some standards and/or contractual requirements may require independent oversight by an external 3rd party such as a Certified Welding Inspector or CWI (AWS), Authorized Inspector or AI (NBIC), or Examiner (ISO).

The qualifying organization is responsible for documenting the procedure qualification in the form of a procedure qualification record (PQR) and certifying the PQR meets all the applicable requirements. Retention of the test coupons and any documentation related to examination and testing is not required after the qualifying organization accepts and documents the results on the PQR.

It is permitted to subcontract any or all the work necessary for preparing the materials to be joined, the subsequent work for preparing test specimens from the completed test joint, and the performance of nondestructive examination and testing, provided the qualifying organization accepts full responsibility for any subcontracted work. While AWS B2.1 does not address simultaneous procedure qualification by multiple organizations, ASME IX has provisions which allow multiple organizations to perform a procedure qualification simultaneously (ref. ASME IX QG-106.4). In order to satisfy such provisions, each participating qualifying organization shall individually have full control and oversight of the qualification.

## 4.6 Weld Development

If you cannot use or modify an existing welding procedure, then you must qualify a new one. As a starting point you may have to perform some weld development and will need guidance on welding parameters and technique. If an existing welding procedure has similar essential variables, you may be able to use it as a starting point. Other sources of reference include information from a filler metal or power supply manufacturer, published papers or books, advice from an experienced welder or colleague, and as a last resort developing the procedure from scratch by trial and error. Each of these are discussed separately in the following sections.

| Weld Layer(s) | Process | Filler Metal Class | Filler Metal Diameter | Current Type / Polarity | Current Amperage Range | Voltage Range | Travel Speed Range |
|---|---|---|---|---|---|---|---|
| Tack | SMAW | E7018 | 3/32 in. [2.5 mm] | DCEP (Reverse) | 75-115 | 18-27 | All |
| Root | SMAW | E7018 | 1/8 in. [3 mm] | DCEP (Reverse) | 85-130 | 18-27 | All |
| Fill & Cap | SMAW | E7018 | 5/32 in. [4 mm] | DCEP (Reverse) | 115-160 | 18-27 | All |

**Fig. 4.12** Example weld parameter table from a SMAW WPS

### 4.6.1 Guidance from an Existing Welding Procedure

In some cases, you may have an existing welding procedure which was developed for a similar application that meets most but not all of the required essential variables. For arc welding processes, most standards require welding parameters such arc voltage and welding current be addressed as non-essential variables so the welding procedure will typically have recommended values listed which can be used for guidance. Examples of welding parameters for SMAW are shown in Fig. 4.12.

### 4.6.2 Guidance from the Filler Metal Manufacturer

Manufacturers of filler metals typically publish suggested welding parameters for their electrodes. These suggestions can be utilized as a starting point, especially if you don't have a lot of experience developing welding procedures or lack a sound understanding of the effect of welding variables. An example of suggested welding parameters from a filler metal manufacturer for GMAW with various electrode diameters and transfer modes are shown in Fig. 4.13.

| Diameter, Polarity Shielding Gas | CTWD in. [mm] | Wire Feed Speed in./min. [m./min.] | Voltage (volts) | Approximate Current (amps) |
|---|---|---|---|---|
| 0.035 in. [0.9 mm], DCEP (Reverse) | | | | |
| 100% $CO_2$ Short Circuit Transfer | 3/8 - 1/2 [9 - 12] | 100 [2.5] | 18 | 80 |
| | | 150 [3.8] | 19 | 120 |
| | | 250 [6.4] | 22 | 175 |
| 90% Ar / 10% $CO_2$ Spray Transfer | 1/2 - 3/4 [12 – 19] | 375 [9.5] | 23 | 195 |
| | | 500 [12.7] | 29 | 230 |
| | | 600 [15.2] | 30 | 275 |
| 0.045 in. [1.1 mm], DCEP (Reverse) | | | | |
| 100% $CO_2$ Short Circuit Transfer | 1/2 - 3/4 [12 – 19] | 125 [3.2] | 19 | 145 |
| | | 150 [3.8] | 20 | 165 |
| | | 200 [5.1] | 21 | 200 |
| 90% Ar / 10% $CO_2$ Spray Transfer | 1/2 - 3/4 [12 – 19] | 350 [8.9] | 27 | 285 |
| | | 475 [12.1] | 30 | 335 |
| | | 500 [12.7] | 30 | 340 |

**Fig. 4.13** Example of suggested GMAW welding parameters for a ER70S-6 solid-wire electrode

### 4.6.3 Guidance from the Welding Power Supply Manufacturer

Manufacturers of welding power supplies typically pre-program welding parameters for certain combinations for filler metals, wire size, shielding gas, etc. This can get you started, especially if you don't have a lot of experience developing welding procedures or lack a sound understanding of the effect of welding variables. An example of the power supply settings for FCAW is shown in Fig. 4.14.

### 4.6.4 Guidance from a Reference

There are many references sources that can provide guidance such as published papers or a book. An example of a reference table from the AWS Welding Handbooks for welding carbon steel with GMAW in spray transfer is shown in Fig. 4.15.

### 4.6.5 Guidance from an Experienced Welder or Colleague

You may work with or know a welder or colleague who has experience working with a particular welding process, material, and/or application who could provide suggestions as a starting point.

| Size in. [mm] | Shielding | Waveform | Gas Type | Mode # | Arc Control | Max. Range |
|---|---|---|---|---|---|---|
| All | Self | CV-SS | N/A | 6 | Pinch | 10 – 45 v |
| All | Gas | CV-GS | Any | 7 | Pinch | 10 – 45 v |
| 0.045 in. [1.1 mm] | Gas | CV | CO2<br>Argon Mix | 90<br>91 | Pinch | 175 – 600 ipm [4.45 – 15.24 m/min]<br>175 – 600 ipm [4.45 – 15.24 m/min] |
| 0.052 in. [1.3 mm] | Gas | CV | CO2<br>Argon Mix | 187<br>188 | Pinch | 150 – 500 ipm [3.81 – 12.7 m/min]<br>150 – 500 ipm [3.81 – 12.7 m/min] |
| 0.063 in. [1.5 mm] | Gas | CV | CO2<br>Argon Mix | 189<br>190 | Pinch | 125 – 400 ipm [3.18 – 10.16 m/min]<br>125 – 375 ipm [3.18 – 10.16 m/min] |

**Fig. 4.14** Example suggested FCAW power supply settings

| Material Thickness | Weld Type | Electrode Diameter | Amperage | Voltage | Wire Feed Speed | Travel Speed |
|---|---|---|---|---|---|---|
| 0.125 in. [3.2 mm] | Butt | 0.035 in. [0.9 mm] | 190 | 26 | 350 in./min. [148 mm/s] | 20 - 25 in./min. [8 - 11 mm/s] |
| 0.250 in. [6.4 mm] | Butt | 0.045 in. [1.1 mm] | 320 | 29 | 400 in./min. [169 mm/s] | 17 - 22 in./min. [7 - 9 mm/s] |
| 0.375 in. [9.5 mm] | Butt | 0.045 in. [1.1 mm] | 300 | 29 | 365 in./min. [154 mm/s] | 11 - 16 in./min. [5 - 7 mm/s] |
| 0.375 in. [9.5 mm] | Fillet | 0.063 in. [1.6 mm] | 300 | 26 | 205 in./min. [87 mm/s] | 10 - 15 in./min. [4 - 6 mm/s] |
| 0.500 in. [12.7 mm] | Butt | 0.063 in. [1.6 mm] | 320 | 26 | 195 in./min.[82 mm/s] | 17 - 22 in./min. [7 - 9 mm/s] |
| 0.750 in. [19.2 mm] | Fillet | 0.063 in. [1.6 mm] | 360 | 27 | 235 in./min. [99 mm/s] | 10 - 15 in./min. [4 - 6 mm/s] |
| Note: DCEP (Reverse polarity), Horizontal or Flat position, 98%Ar-2%$O_2$ shielding gas at 40-50 cfh. | | | | | | |

**Fig. 4.15** Example suggested GMAW welding parameters from the AWS Welding Handbooks

### 4.6.6 Trial and Error

Ideally you should never have to start from scratch but in some cases you may be welding materials that are not common, have to account for variables that are not typical, or are limited by available welding equipment. If this is the route you must take it is imperative that you perform some kind of testing! It may be a simulated test where the test weldment is placed into service to evaluate its performance. It may be destructive testing such as performing a macro-etch to evaluation the microstructure or bend tests such as those required when performing procedure qualifications. You should never just use the outward appearance of the weld bead as your only gauge of quality, especially when you are working with limitations in terms of welding equipment or welder skill.

### 4.6.7 Welding Procedure Optimization

Different parts of the weldment may require different welding parameters, techniques, and sequence. Some weld development and optimization may be required depending on many factors such as the welding process, weld joint configuration, welding position, alloys etc. For example, the welding parameters and technique you utilize for a V-groove's root pass and hot pass may be different than its fill passes and cap passes. For a V-groove as illustrated in Fig. 4.16, the root and hot passes (layers 1 and 2) may be string beads, followed by a fill pass with weaving producing a single bead for layer 3, followed by layer 4 (beads 4 and 5) produced by multiple beads (aka "split beads"), followed by several cap passes (beads 6 and 7) also produced by multiple beads. Additionally, the root gap and root land dimensions of the joint may need optimized in order to ensure good penetration of the root pass.

If the joint was a V-groove in a pipe and the weld is being produced orbital 5G or 6G, the welding parameters may also have to be altered as each weld progresses around the joint in order to account for gravity to ensure a sound weld is being deposited. For a horizontal 2G groove weld in a pipe, the bead placement and progression also has to be considered in order to account for gravity so that no lack-of-fusion defects are produced (see Fig. 4.17).

In addition to trial-and-error methods, there are numerous methods to develop mathematical relationships between the welding process input parameters and the output

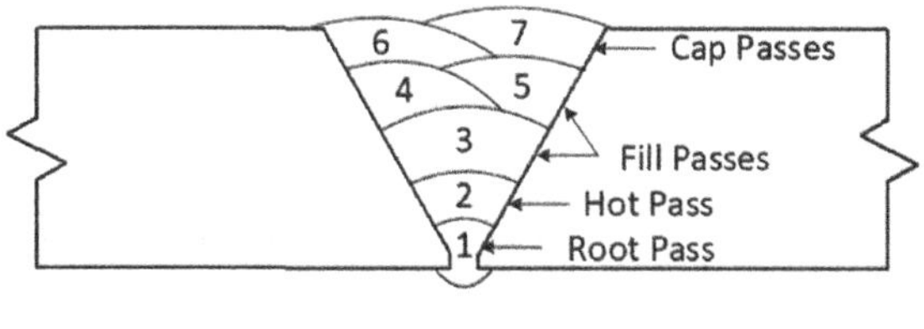


**Fig. 4.16** V-groove illustrating the various weld layers and bead placement

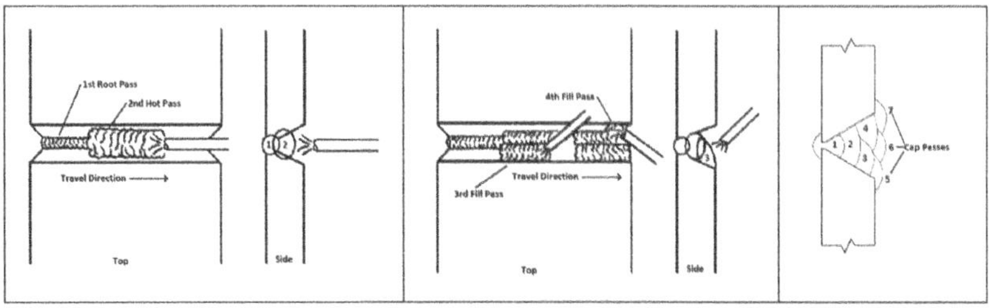


**Fig. 4.17** Example of welding technique and bead placement for a horizontal 2G pipe weld

variables in order to determine the influence of the input parameters that lead to the desired weld properties. Some of these methods include:

- Design-of-Experiments (DOE)
- Response Surface Methodology (RSM)
- Taguchi Methods
- Neural Network Modelling
- Numerical Modelling.

An example process diagram for an arc welding process is shown in Fig. 4.18 with 4 process input variables and 1 process output variables. Process diagrams are fundamental to understanding and developing mathematical relationships between the welding process input parameters and the output response variables.

Another method for optimizing welding procedures called ARCWISE was developed by EWI.[5] The sequence of testing as follows produce parametric relationships such as shown in Fig. 4.19.

- Systematic parameter development method
- Weld bead sizing and acceptance criteria
- Constant deposit area tests
- Constant arc length testing

**Fig. 4.18** Diagram of an arc welding process showing its input/output relationships

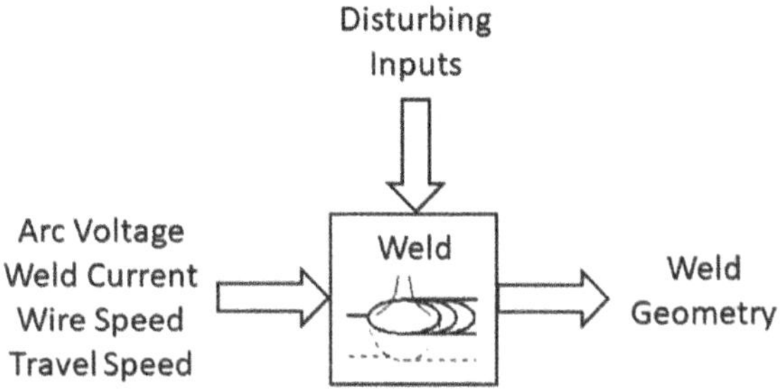

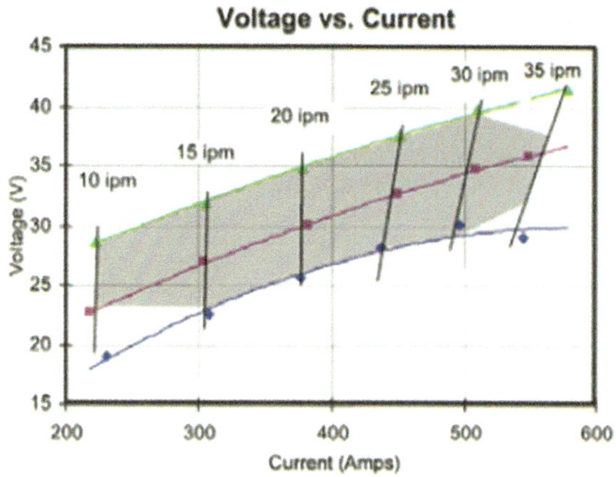


**Fig. 4.19** Parametric relationships developed by the ARCWISE method[5]

- Parametric data acquisition and analysis
- Bead shape analysis
- Welding productivity determination.

## 4.7 Preliminary Welding Procedure Specification (pWPS)

A preliminary welding procedure or pWPS is simply a draft welding procedure that is utilized during the procedure qualification process to guide the welder/operator in completing the qualification test coupon. Drafting a pWPS is not required by most qualification standards, however it is good practice as it provides the welding engineer an opportunity to summarize and review all the requirements for the procedure qualification test while ensuring the welder/operator performs the test correctly. The format of a pWPS is typically the same format utilized for the qualified WPS, addressing all essential, supplementary essential, and non-essential variables for the given welding process.

Note that the acronym for a preliminary welding procedure is preceded by a small case letter "p" versus the acronym for a prequalified welding procedure which is preceded by a capital letter "P" i.e. pWPS versus PWPS.

## 4.8 Qualification Test Weldment

The qualification test weldment should be welded by a be welded by a qualified welder or welding operator with oversight from the welding engineer. All required essential, supplementary essential if applicable, and non-essential variables should be documented. Depending on the type of WPS being qualified, it's also a good idea to record the actual preheat/interpass temperatures and welding parameters for each bead/run on a "bead log",

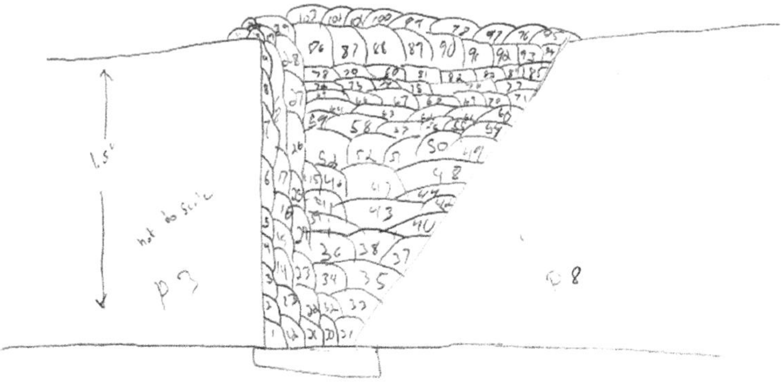

**Fig. 4.20** Example weld bead map

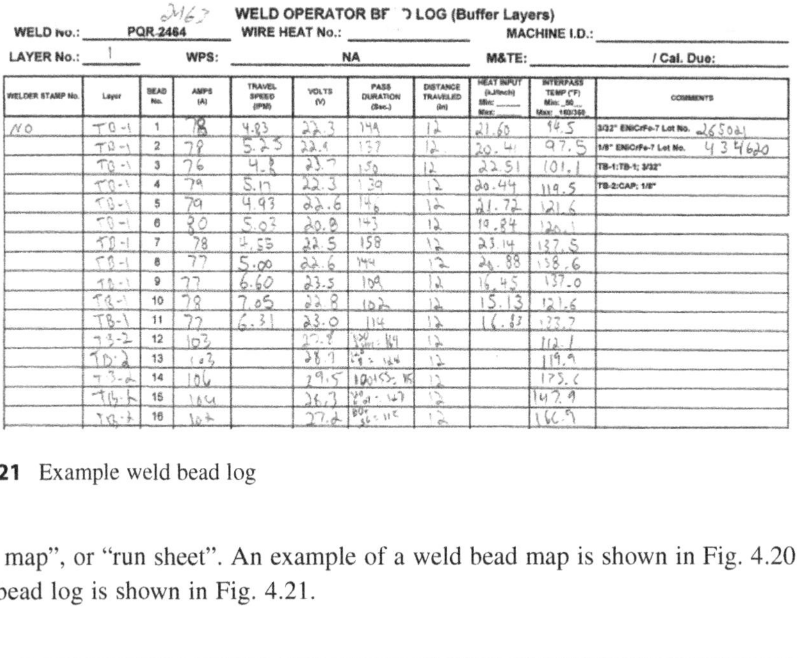

WELD OPERATOR BE[illegible]D LOG (Buffer Layers)

WELD No.: PQR 2464 WIRE HEAT No.: MACHINE I.D.:

LAYER No.: 1 WPS: NA M&TE: / Cal. Due:

| WELDER STAMP No. | Layer | BEAD No. | AMPS (A) | TRAVEL SPEED (IPM) | VOLTS (V) | PASS DURATION (Sec.) | DISTANCE TRAVELED (in) | HEAT INPUT (kJ/inch) Min: Max: | INTERPASS TEMP (°F) Min: 50 Max: 160/350 | COMMENTS |
|---|---|---|---|---|---|---|---|---|---|---|
| NO | TB-1 | 1 | 78 | 4.83 | 22.3 | 149 | 12 | 21.60 | 94.5 | 3/32" ENiCrFe-7 Lot No. 265021 |
| | TB-1 | 2 | 78 | 5.25 | 22.4 | 137 | 12 | 20.41 | 97.5 | 1/8" ENiCrFe-7 Lot No. 434620 |
| | TB-1 | 3 | 76 | 4.8 | 23.7 | 150 | 12 | 22.51 | 101.1 | TB-1:TB-1; 3/32" |
| | TB-1 | 4 | 79 | 5.17 | 22.3 | 139 | 12 | 20.44 | 119.5 | TB-2:CAP; 1/8" |
| | TB-1 | 5 | 79 | 4.93 | 22.6 | 146 | 12 | 21.72 | 121.6 | |
| | TB-1 | 6 | 80 | 5.03 | 20.8 | 143 | 12 | 19.84 | 120.1 | |
| | TB-1 | 7 | 78 | 4.55 | 22.5 | 158 | 12 | 23.14 | 137.5 | |
| | TB-1 | 8 | 77 | 5.00 | 22.6 | 144 | 12 | 20.88 | 138.6 | |
| | TB-1 | 9 | 77 | 6.60 | 23.5 | 109 | 12 | 16.45 | 137.0 | |
| | TB-1 | 10 | 78 | 7.05 | 22.8 | 102 | 12 | 15.13 | 121.6 | |
| | TB-1 | 11 | 77 | 6.31 | 23.0 | 114 | 12 | 16.83 | 123.7 | |
| | TB-2 | 12 | 103 | | 27.8 | [illegible] | 12 | | 112.1 | |
| | TB-2 | 13 | 103 | | 28.7 | [illegible] | 12 | | 119.9 | |
| | TB-2 | 14 | 106 | | 29.5 | [illegible] | 12 | | 175.6 | |
| | TB-2 | 15 | 104 | | 26.3 | [illegible] | 12 | | 147.9 | |
| | TB-2 | 16 | 10[illegible] | | 27.2 | [illegible] | 12 | | 166.9 | |

**Fig. 4.21** Example weld bead log

"bead map", or "run sheet". An example of a weld bead map is shown in Fig. 4.20 and a weld bead log is shown in Fig. 4.21.

## 4.9 Examination and Testing

Once the qualification test weldment is completed, the next step is to perform the required non-destructive examinations followed by destructive testing. Depending on the qualification standard, there may or may not be provisions for retesting if the qualification test weldment fails any of the test acceptance criteria. AWS B2.1 does not have provisions for re-testing, therefore you must re-run the PQR test weldment if a test fails to meet the acceptance criterial. ASME IX does have provisions for re-testing if it can be determined that the cause of the failure was not related to welding parameters (ref. ASME IX QW-202.1). If adequate material of the original test coupon exists, additional test specimens may be removed as close as practicable to the original test location to replace the failed test specimen, or alternatively, another test coupon can be welded using identical welding parameters. All of the testing can be performed in-house by qualified personnel using properly calibrated equipment and procedures, or the testing can be performed externally by a qualified testing laboratory.

## 4.10 Procedure Qualification Documentation

Once the qualification test weldment has met all the examination and testing acceptance requirements, the welding procedure is now deemed qualified and it's time to document the welding procedure specification (WPS) and its corresponding procedure qualification record (PQR). The PQR must address all essential/supplementary essential variables for the welding process and application and include all supporting documentation such as material test reports (base material and filler metal), PWHT records, non-destructive/destructive test reports, and weld bead logs. The WPS must address all essential/supplementary essential/non-essential WPS variables for the welding process and application and may include weld technique sheets (if needed). See Chap. 3 for a more thorough description of all the welding related documentation.

## 4.11 Documentation Audit

An audit is defined as a planned and documented activity performed to determine by investigation, examination, or evaluation of objective evidence the adequacy of and compliance with established procedures, instructions, drawings, and other applicable documents, and the effectiveness of implementation. An audit of a qualified welding procedure and its associated paperwork ensures that the documentation is in compliance and meets all the requirements the applicable qualification standard(s). The audit should be performed independently by a qualified peer who has a working knowledge of the applicable qualification standard(s). During the audit, a checklist of the welding variables can be used to determine if all essential, supplementary essential, and nonessential variables have been addressed by the Welding Procedure Specification and accompanying Procedure Qualification Record. An example of an audit checklist for a Shielded Metal Arc Hardfacing/Corrosion Resistant Overlay WPS and PQR are shown in Fig. 4.22. A hard-copy of the audit is normally signed and filed along with the corresponding WPS and PQR for record.

**Shielded Metal Arc Welding (SMAW) Hardfacing/Overlay WPS/PQR Audit Form**

**ASME BPVC IX 2021ed**

| | |
|---|---|
| WPS # Under Review: | Dated: |
| Supporting PQR(s) #: | Dated: |

| Paragraph | | Brief of Variables | HFO Ess. | CRO Ess. | Non-Ess. | WPS | PQR |
|---|---|---|---|---|---|---|---|
| QW-402 Joints | .16 | < Finished *t* | X | X | | | |
| QW-403 Base Metals | .20 | ø P-Number | X | X | | | |
| | .23 | ø *T* Qualified | X | X | | | |
| QW-404 Filler Metals | .12 | ø Classification | X | | | | |
| | .37 | ø A-Number | | X | | | |
| | .38 | ø Diameter | | | X | | |
| QW-405 Positions | .4 | + Position | X | X | | | |
| QW-406 Preheat | .4 | Preheat Decrease > 100°F (Δ55°F) | X | X | | | |
| | .4 | > Interpass | X | X | | | |
| QW-407 PWHT | .6 | ø PWHT | X | | | | |
| | .9 | ø PWHT | | X | | | |
| QW-409 Electrical | .4 | ø Current or polarity | X | X | | | |
| | .8 | ø I & E range | | | X | | |
| | .22 | 1st layer - Inc. > 10% | X | X | | | |
| QW-410 Technique | .1 | ø String / Weave | | | X | | |
| | .5 | ø Method cleaning | | | X | | |
| | .26 | ± Peening | | | X | | |
| | .38 | ø Mutliple to single layer | X | X | | | |
| QW-453 Testing | | PT Testing per QW-195 | X | X | | | |
| | | 4 Bend Tests per QW-160 | | X | | | |
| | | 3 Hardness Readings per QW-462 | X | | | | |
| | | Visual Examination per QW-462 | X | | | | |
| | | Chemical Analysis (Opt. per QW-462) | X | | | | |

**Additional Requirements:**

**Reviewer Comments:**

**Fig. 4.22** Example SMAW HFO/CRO audit form

## 4.12 Summary

Obtaining a qualified welding procedure can be as easy as taking advantage of a Standard Welding Procedure Specification (SWPS) or Prequalified Welding Procedure Specification (PWPS), or as complicated as having to perform weld development then producing a qualification test weldment to meet multiple standards with numerous testing requirements. Some companies may only qualify a couple of welding procedures a year while others may have to qualify numerous procedures. Regardless, any company who works in an industry that requires welding procedures meet an applicable standard should have well documented welding/quality programs and procedures which has requirements for document retention.

**Notes**

1. 2001, Survey of Joining, Cutting, and Allied Processes, Welding Handbook, Vol. 1: Welding Science and Technology, American Welding Society.
2. 2021, Sturgill, P.L., Options for Qualifying Welding Procedure Specifications, Inspection Trends, 24(1):17–19, American Welding Society.
3. 2018, Song, W. and Liu, X., "Fatigue assessment of steel load-carrying cruciform welded joints by means of local approaches", Fatigue and Fracture of Engineering Materials and Structures, Wiley, 41(12):2598–2613.
4. 2019, Garašić, I., et al., "Determination of Ballistic Properties of ARMOX 500 T Steel Welded Joint", Engineering Review, 39(2):186–196.
5. 2000, Harwig, D.D., "A wise method for assessing arc welding performance and quality", Welding Journal 79(12):35–39. American Welding Society.

# 5 Procedure Qualification for Groove and Fillet Welds

The concepts of developing a qualified welding procedure and producing the required welding documentation was introduced in Chap. 4. This chapter discusses the specific requirements for the qualification of a groove or fillet welding procedure. The required documented welding variables for the Procedure Qualification Record (PQR) and their qualification limits for the Welding Procedure Specification (WPS) are discussed, along with the required examination and test methods and their acceptance criteria are outlined.

It should be noted that both AWS B2.1 and ASME IX allow qualification of a fillet welding procedure with a groove weld configuration, but not vice versa (ref. AWS B2.1 4.5.3 and 4.8.1 and ASME IX QW-202.2(c)). A limitation of ASME IX states that pressure retaining fillet welds must be qualified with a groove weld test coupon using groove weld testing requirements, but non-pressure retaining fillet welds can be qualified with a fillet weld test coupon.

## 5.1 Groove and Fillet Welding Variables

The following tables provide the required welding variables for fillet and groove welding procedures qualified under AWS B2.1 (ref. Tables 4.1.3 and 4.1.4) and ASME IX (ref. Tables QW-253, QW-254, QW-255, and QW-256 and Article IV). For convenience, the variables are grouped as discussed in Sect. 4.1.9. The Essential (E), Supplementary Essential (SE), and Non-Essential (NE) variables are compared for the five most common arc welding procedures (SMAW, GMAW, FCAW, GTAW, SAW). Variables marked N/A are not addressed for a given standard. The tables should be regarded as illustrative only since the various standards are revised periodically.

© The Author(s), under exclusive license to Springer Nature Switzerland AG 2025

D. Barborak, *Arc Welding Qualification Standards*, Synthesis Lectures on Welding Engineering, https://doi.org/10.1007/978-3-031-64646-1_5

### 5.1.1 Joint Design

See Table 5.1.

### 5.1.2 Base Metal

See Table 5.2.

### 5.1.3 Filler Metals

See Table 5.3.

### 5.1.4 Position

See Table 5.4.

### 5.1.5 Preheat and Interpass Temperature

See Table 5.5.

### 5.1.6 Postweld Heat Treatment

See Table 5.6.

### 5.1.7 Shielding Gas

See Table 5.7.

### 5.1.8 Electrical Characteristics

See Table 5.8.

### 5.1.9 Other Variables

See Table 5.9.

**Table 5.1** Joint design variables for a groove or fillet welding procedure

| AWS B2.1 | | | | | | | ASME IX | | | | | | |
|---|---|---|---|---|---|---|---|---|---|---|---|---|---|
| Variable | Ref. | SMAW | GMAW | FCAW | GTAW | SAW | Variable | Ref. | SMAW | GMAW | FCAW | GTAW | SAW |
| Joint type and dimensions | 4.13.1(1) | NE | NE | NE | NE | NE | A change in the type of groove (V-groove, U-groove, single-bevel, double-bevel, etc.) | QW-402.1 | NE | NE | NE | NE | NE |
| A change from a fillet to a groove weld | 4.14.1(2) | E | E | E | E | E | N/A | | | | | | |
| N/A | | | | | | | A change in the specified root spacing | QW-402.10 | NE | NE | NE | NE | NE |
| Treatment of backside, method of gouging preparation | 4.13.1(2) | NE | NE | NE | NE | NE | N/A | | | | | | |
| Backing material if used | 4.13.1(3) | NE | NE | NE | NE | NE | The deletion of the backing in single welded groove welds. Double-welded groove welds are considered welding with backing | QW-402.4 | NE | NE | NE | | NE |
| A change in M-Number of backing | 4.14.1(3) | E | E | E | E | E | The addition of a backing or a change in its nominal composition | QW-402.5 | | | | NE | |
| The addition of thermal back gouging on M-11, M-23, M-24, M-26, or M-27 heat-treatable base metal | 4.14.1(4) | E | E | E | E | E | N/A | | | | | | |
| N/A | | | | | | | The addition or deletion of nonmetallic retainers or nonfusing metal retainers | QW-402.11 | NE | NE | NE | NE | NE |

**Table 5.2** Base metal variables for a groove or fillet welding procedure

| AWS B2.1 | | | | | | | ASME IX | | | | | | |
|---|---|---|---|---|---|---|---|---|---|---|---|---|---|
| Variable | Ref. | SMAW | GMAW | FCAW | GTAW | SAW | Variable | Ref. | SMAW | GMAW | FCAW | GTAW | SAW |
| M-number and group number | 4.13.2(1) | NE | NE | NE | NE | NE | N/A | | | | | | |
| A change from one M-number base metal to another M-number base metal or to a combination of M-number base metals except as permitted in 4.3.10 | 4.14.2(2) | E | E | E | E | E | Base metals specified in the WPS shall be qualified by a procedure qualification test that was made using base metals in accordance with QW-424 | QW-403.11 | E | E | E | E | E |
| A change from one unlisted base metal to a different unlisted base metal or to a listed base metal or vice versa | 4.14.2(3) | E | E | E | E | E | N/A | | | | | | |
| A change from one M-number group to any other M-number group number, except as permitted in 4.3.11 | 4.14.2(4) | SE | SE | SE | SE | SE | Welding procedure specifications shall be qualified using one of the following:<br>(a) the same base metal (including type or grade) to be used in production welding<br>(b) for ferrous materials, a base metal listed in the same P-number group number in Table QW/QB-422 as the base metal to be used in production welding | QW-403.5 | SE | SE | SE | SE | SE |

(continued)

**Table 5.2** (continued)

| AWS B2.1 | | | | | | | ASME IX | | | | | | |
|---|---|---|---|---|---|---|---|---|---|---|---|---|---|
| Variable | Ref. | SMAW | GMAW | FCAW | GTAW | SAW | Variable | Ref. | SMAW | GMAW | FCAW | GTAW | SAW |
| | | | | | | | (c) for nonferrous materials, a base metal listed with the same P-number UNS number in Table QW/QB-422 as the base metal to be used in production welding<br>For ferrous materials in Table QW/QB-422, a procedure qualification shall be made for each P-number group number combination of base metals, even though procedure qualification tests have been made for each of the two base metals welded to itself. If, however, two or more qualification records have the same essential and supplementary essential variables, except that the base metals are assigned to different group numbers within the same P-number, then the combination of base metals is also qualified. In addition, when base metals of two different group numbers within the same P-number are qualified using a single test coupon, that coupon qualifies the welding of those two group numbers within the same P-number to themselves as well as to each other using the variables qualified | | | | | | |

(continued)

**Table 5.2** (continued)

| AWS B2.1 | | | | | | | ASME IX | | | | | | |
|---|---|---|---|---|---|---|---|---|---|---|---|---|---|
| Variable | Ref. | SMAW | GMAW | FCAW | GTAW | SAW | Variable | Ref. | SMAW | GMAW | FCAW | GTAW | SAW |
| A change from one M-5 group (A, B, etc.) to any other. A change from M-9A to M-9B, but not vice versa. A change from M-10 to M-11 group (A, B, etc.) to any other group | 4.14.2(5) | E | E | E | E | E | N/A | | | | | | |
| Thickness range qualified | 4.13.2(2) | NE | NE | NE | NE | NE | A change in base metal thickness beyond the range qualified in QW-451, except as otherwise permitted by QW-202.4(b) | QW-403.8 | E | E | E | E | E |
| A change in base metal thickness beyond the range permitted in 4.5 | 4.14.2(1) | E | E | E | E | E | | | | | | | |
| N/A | | | | | | | For single-pass or multipass welding in which any pass is greater than 1/2 in. (13 mm) thick, an increase in base metal thickness beyond 1.1 times that of the qualification test coupon | QW-403.9 | E | E | E | | E |
| The coating description or type, if present | 4.13.2(4) | NE | NE | NE | NE | NE | N/A | | | | | | |
| A change from an uncoated metal to a metal coated (such as painted or galvanized) metal unless the coating is removed from the weld area prior to welding, but not vice versa, as permitted in 4.3.8 | 4.14.2(6) | E | E | E | E | E | | | | | | | |

**Table 5.3** Filler metals variables for a groove or fillet welding procedure

| AWS B2.1 | | | | | | | ASME IX | | | | | | |
|---|---|---|---|---|---|---|---|---|---|---|---|---|---|
| Variable | Ref. | SMAW | GMAW | FCAW | GTAW | SAW | Variable | Ref. | SMAW | GMAW | FCAW | GTAW | SAW |
| Specification, classification, F- and A-Number, or if not classified the nominal composition | 4.13.1(1) | NE | NE | NE | NE | NE | Implied by QW-404.4, QW-404.5, QW-404.12, & QW-404.33 | | | | | | |
| A change from one F-number to any other F-number or to any filler metal not listed in Annex B | 4.14.3(1) | E | E | E | E | E | A change from one F-number in Table QW-432 to any other F-number or to any other filler metal not listed in Table QW-432 | QW-404.4 | E | E | E | E | E |
| For ferrous materials, a change from one A-number to any other A-number or to a filler metal analysis not listed in Annex B (the PQR and WPS shall state the nominal chemical composition, the AWS specification and classification, or the manufacturer's designation for filler metals which do not fall in an A-number group). Qualification with A-1 shall qualify for A-2 and vice versa | 4.14.3(2) | E | E | E | E | E | (Applicable only to ferrous metals.) A change in the chemical composition of the weld deposit from one A-number to any other A-number in Table QW-442. Qualification with A-No. 1 shall qualify for A-No. 2 and vice versa | QW-404.5 | E | E | E | E | E |
| A change in AWS filler metal specification and classification | 4.14.3(4) | SE | SE | SE | SE | SE | A change in the filler metal classification within an SFA specification, or for a filler metal not covered by an SFA specification or a filler metal with a "G" suffix within an SFA specification, a change in the trade name of the filler metal | QW-404.12 | SE | SE | SE | SE | |

(continued)

**Table 5.3** (continued)

| AWS B2.1 | | | | | | | ASME IX | | | | | | |
|---|---|---|---|---|---|---|---|---|---|---|---|---|---|
| Variable | Ref. | SMAW | GMAW | FCAW | GTAW | SAW | Variable | Ref. | SMAW | GMAW | FCAW | GTAW | SAW |
| N/A | | | | | | | A change in the filler metal classification within an SFA specification, or, if not conforming to a filler metal classification within an SFA specification, a change in the manufacturer's trade name for the filler metal. When optional supplemental designators, such as those which indicate moisture resistance (i.e., XXXXR), diffusible hydrogen (i.e., XXXX H16, H8, etc.), and supplemental toughness testing (i.e., XXXX-1 or EXXXXM), are specified on the WPS, only filler metals which conform to the classification with the optional supplemental designator(s) specified on the WPS shall be used | QW-404.33 | NE | NE | NE | NE | NE |
| A change from wire to strip electrodes and vice versa | 4.14.3(15) | | | | | E | N/A | | | | | | |
| Weld metal thickness by process and filler metal classification | 4.13.3(2) | NE | NE | NE | NE | NE | N/A | | | | | | |

(continued)

**Table 5.3** (continued)

| AWS B2.1 | | | | | | | ASME IX | | | | | | |
|---|---|---|---|---|---|---|---|---|---|---|---|---|---|
| Variable | Ref. | SMAW | GMAW | FCAW | GTAW | SAW | Variable | Ref. | SMAW | GMAW | FCAW | GTAW | SAW |
| A change in the weld metal thickness beyond that permitted in 4.5 | 4.14.3(17) | E | E | E | E | E | A change in deposited weld metal thickness beyond that qualified in accordance with QW-451 for procedure qualification or QW-452 for performance qualification, except as otherwise permitted in QW-303.1 and QW-303.2. When a welder is qualified using volumetric examination, the maximum thickness stated in Table QW-452.1(b) applies | QW-404.30 | E | E | E | E | E |
| Filler metal size or diameter | 4.14.3(3) | NE | NE | NE | NE | NE | A change in the nominal size of the electrode or electrodes specified in the WPS | QW-404.6 | NE | NE | NE | | NE |
| | | | | | | | A change in the size of the filler metal | QW-404.3 | | | | NE | |
| Flux-electrode classification | 4.13.3(4) | | | | | NE | (a) A change in the indicator for minimum tensile strength (e.g., the 7 in F7A2-M12K) when the flux wire combination is classified in Section II, Part C<br>(b) A change in either the flux trade name or wire trade name when neither the flux nor the wire is classified in Section II, Part C<br>(c) A change in the flux trade name when the wire is classified in Section II, Part C but the flux is not classified. A change in the wire classification within the requirements of QW-404.5 does not require requalification<br>(d) A change in the flux trade name for A-No. 8 deposits | QW-404.9 | | | | | E |

(continued)

**Table 5.3** (continued)

| AWS B2.1 | | | | | | | ASME IX | | | | | | |
|---|---|---|---|---|---|---|---|---|---|---|---|---|---|
| Variable | Ref. | SMAW | GMAW | FCAW | GTAW | SAW | Variable | Ref. | SMAW | GMAW | FCAW | GTAW | SAW |
| N/A | | | | | | | A change in the flux trade name and designation | QW-404.29 | | | | | NE |
| N/A | | | | | | | A change in flux type (i.e., neutral to active or vice versa) for multilayer deposits in P-No. 1 materials | QW-404.34 | | | | | E |
| If the weld metal alloy content is largely dependent upon the composition of the flux, any change in the welding procedure which would result in the important weld metal alloying elements being outside the specified chemical composition range of the WPS | 4.14.3(6) | | | | | E | Where the alloy content of the weld metal is largely dependent upon the composition of the flux used, any change in any part of the welding procedure which would result in the important alloying elements in the weld metal being outside of the specification range of chemistry given in the welding procedure specification | QW-404.10 | | | | | E |
| A change from one AWS flux-electrode classification listed to any other electrode flux-electrode classification, or to an unlisted electrode-flux classification. A variation of 0.5% of the molybdenum content of the weld metal does not require requalification | 4.14.3(16) | | | | | E | A change in the flux-wire classification or a change in either the electrode or flux trade name when the flux-wire combination is not classified to an SFA specification. Requalification is not required when a flux-wire combination conforms to an SFA specification and the change in classification is (a) from one diffusible hydrogen level to another (e.g., a change from F7A2-EA1-A1-H4 to F7A2-EA1-A1-H16), or (b) to a larger number in the indicator for toughness, indicating classification at a lower toughness testing temperature (e.g., a change from F7A2-EM12K to F7A4-EM12K) | QW-404.35 | | | | | SE |

(continued)

**Table 5.3** (continued)

| AWS B2.1 | | | | | | | ASME IX | | | | | | |
|---|---|---|---|---|---|---|---|---|---|---|---|---|---|
| Variable | Ref. | SMAW | GMAW | FCAW | GTAW | SAW | Variable | Ref. | SMAW | GMAW | FCAW | GTAW | SAW |
| N/A | | | | | | | Recrushed slag: When flux from recrushed slag is used, each batch or blend, as defined in SFA-5.01, shall be tested in accordance with Section II, Part C by either the manufacturer or the user | QW-404.36 | | | | | E |
| Penetration enhancing flux | 4.13.3(5) | | | | NE | | The addition or deletion of flux to the face of a weld joint for the purpose of affecting weld penetration | QW-404.50 | | | | NE | |
| Supplemental filler metal | 4.13.3(6) | NE | NE | NE | NE | NE | The addition, deletion, or change of more than 10% in the volume of supplemental filler metal | QW-404.24 | | E | E | | E |
| Addition or deletion of supplementary filler metal (powder or wire), or a change of 10% in the amount | 4.14.3(11) | E | E | SE | SE | E | | | | | | | |
| Addition or deletion, or a change in the nominal amount or composition of supplementary metal (in addition to filler metal) beyond that qualified | 4.14.3(14) | | E | | | E | Where the alloy content of the weld metal is largely dependent upon the composition of the supplemental filler metal (including powder filler metal for PAW), any change in any part of the welding procedure that would result in the important alloying elements in the weld metal being outside of the specification range of chemistry given in the welding procedure specification | QW-404.27 | | E | E | | E |

(continued)

**Table 5.3** (continued)

| AWS B2.1 | | | | | | | ASME IX | | | | | | |
|---|---|---|---|---|---|---|---|---|---|---|---|---|---|
| Variable | Ref. | SMAW | GMAW | FCAW | GTAW | SAW | Variable | Ref. | SMAW | GMAW | FCAW | GTAW | SAW |
| The addition or deletion of filler material | 4.14.3(8) | | | | E | | The deletion or addition of filler metal | QW-404.14 | | | | E | |
| Consumable insert and type | 4.13.3(7) | | | | NE | | The omission or addition of consumable inserts. Qualification in a single-welded butt joint, with or without consumable inserts, qualifies for fillet welds and single-welded butt joints with backing or double-welded butt joints. Consumable inserts that conform to SFA-5.30, except that the chemical analysis of the insert conforms to an analysis for any bare wire given in any SFA specification or AWS Classification, shall be considered as having the same F-number as that bare wire as given in Table QW-432 | QW-404.22 | | | | NE | |
| Energized filler metal "hot" | 4.13.3(10) | | | | NE | | N/A | | | | | | |
| N/A | | | | | | | A change from one of the following filler metal product forms to another:<br>(a) bare (solid or metal cored)<br>(b) flux cored<br>(c) flux coated (solid or metal cored)<br>(d) powder | QW-404.23 | | E | E | E | |

**Table 5.4** Position variables for a groove or fillet welding procedure

| AWS B2.1 | | | | | | | ASME IX | | | | | | |
|---|---|---|---|---|---|---|---|---|---|---|---|---|---|
| Variable | Ref. | SMAW | GMAW | FCAW | GTAW | SAW | Variable | Ref. | SMAW | GMAW | FCAW | GTAW | SAW |
| Welding positions(s) | 4.13.4(1) | NE | NE | NE | NE | NE | The addition of other welding positions than those already qualified. see QW-120, QW-130, QW-203, and QW-303 | QW-405.1 | NE | NE | NE | NE | NE |
| Progression for vertical welding | 4.13.4(2) | NE | NE | NE | NE | NE | A change from upward to downward, or from downward to upward, in the progression specified for any pass of a vertical weld, except that the cover or wash pass may be up or down. The root pass may also be run either up or down when the root pass is removed to sound weld metal in the preparation for welding the second side | QW-405.3 | NE | NE | NE | NE | NE |
| A change from any position to the vertical position, uphill progression. Vertical uphill progression qualifies all positions | 4.14.4(1) | SE | SE | SE | SE | | N/A | | | | | | |

**Table 5.5** Preheat and interpass temperature variables for a groove or fillet welding procedure

| AWS B2.1 | | | | | | | ASME IX | | | | | | |
|---|---|---|---|---|---|---|---|---|---|---|---|---|---|
| Variable | Ref. | SMAW | GMAW | FCAW | GTAW | SAW | Variable | Ref. | SMAW | GMAW | FCAW | GTAW | SAW |
| Preheat minimum | 4.13.5(1) | NE | NE | NE | NE | NE | A decrease of more than 100 °F (55 °C) in the preheat temperature qualified. The minimum temperature for welding shall be specified in the WPS | QW-406.1 | E | E | E | E | E |
| A decrease in preheat of more than 100 °F [38 °C] from that qualified | 4.14.5(1) | E | E | E | E | E | | | | | | | |
| Interpass temperature maximum (if applicable) | 4.13.5(2) | NE | NE | NE | NE | NE | An increase of more than 100°F (55 °C) in the maximum interpass temperature recorded on the PQR. This variable does not apply for any of the following conditions:<br>(a) WPS is qualified with a heat treatment above the upper transformation temperature<br>(b) WPS is for welding austenitic or P-10H material and is qualified with a solution heat treatment<br>(c) Base metals are assigned to P-No. 8, P-Nos. 21 through 26, and P-Nos. 41 through 49 | QW-406.3 | SE | SE | SE | SE | SE |
| An increase of more than 100 °F [38 °C] in the maximum interpass temperature from that recorded on the PQR | | SE | SE | SE | SE | SE | | | | | | | |
| Preheat maintenance | 4.13.5(3) | NE | NE | NE | NE | NE | A change in the maintenance or reduction of preheat upon completion of welding prior to any required postweld heat treatment | QW-406.2 | NE | NE | NE | | NE |
| For M-23, M-24, M-26, and M-27 heat-treatable materials an increase in the preheat or interpass temperature of more than 100 °F [38 °C] from that qualified | | E | E | | E | | N/A | | | | | | |

**Table 5.6** Postweld heat treatment variables for a groove or fillet welding procedure

| AWS B2.1 | | | | | | | ASME IX | | | | | | |
|---|---|---|---|---|---|---|---|---|---|---|---|---|---|
| Variable | Ref. | SMAW | GMAW | FCAW | GTAW | SAW | Variable | Ref. | SMAW | GMAW | FCAW | GTAW | SAW |
| PWHT temperature and time | 4.13.6(1) | NE | NE | NE | NE | NE | | | | | | | |
| For the following M-Numbers 1, 3, 4, 5, 6, 7, 9, 10, 11, and 12, a change from any one condition to any other requires requalification:<br>(a) No PWHT<br>(b) PWHT below the lower transformation temperature<br>(c) PWHT within the transformation temperature range<br>(d) PWHT above the upper transformation temperature<br>(e) PWHT above the upper transformation temperature, followed by treatment below the lower transformation temperature | 4.14.6(1) | E | E | E | E | E | A separate procedure qualification is required for each of the following:<br>(a) For P-Numbers 1 through 6 and 9 through 15F materials, the following postweld heat treatment conditions apply:<br>(1) no PWHT<br>(2) PWHT below the lower transformation temperature<br>(3) PWHT above the upper transformation temperature (e.g., normalizing)<br>(4) PWHT above the upper transformation temperature followed by heat treatment below the lower transformation temperature (e.g., normalizing or quenching followed by tempering) | QW-407.1 | E | E | E | E | E |
| For all materials not covered above, a separate PQR is required for no PWHT and PWHT | 4.14.6(2) | E | E | E | E | E | N/A | | | | | | |

(continued)

**Table 5.6** (continued)

| AWS B2.1 | | | | | | | ASME IX | | | | | | |
|---|---|---|---|---|---|---|---|---|---|---|---|---|---|
| Variable | Ref. | SMAW | GMAW | FCAW | GTAW | SAW | Variable | Ref. | SMAW | GMAW | FCAW | GTAW | SAW |
| The qualification test weldment shall be subjected to heat treatment essentially equivalent to that of the production weldment, including at least 80% of the aggregate time at temperature | 4.14.6(3) | SE | SE | SE | SE | SE | A change in the postweld heat treatment (see QW-407.1) temperature and time range. The procedure qualification test shall *be subjected to PWHT essentially* equivalent to that encountered in the fabrication of production welds, including at least 80% of the aggregate times at temperature(s). The PWHT total time(s) at temperature(s) may be applied in one heating cycle. This variable does not apply to a WPS qualified for welding base metals that are assigned to P-No. 8, P-Nos. 21 through 26, and P-Nos. 41 through 49 | QW-407.2 | SE | SE | SE | SE | SE |

**Table 5.7** Shielding gas variables for a groove or fillet welding procedure

| AWS B2.1 | | | | | | | ASME IX | | | | | | |
|---|---|---|---|---|---|---|---|---|---|---|---|---|---|
| Variable | Ref. | SMAW | GMAW | FCAW | GTAW | SAW | Variable | Ref. | SMAW | GMAW | FCAW | GTAW | SAW |
| Torch shielding gas and flow rate range | 4.13.7(1) | | NE | NE | NE | | A change in the specified flow rate range of the shielding gas or mixture of gases | QW-408.3 | | NE | NE | | |
| Root shielding gas and flow rate range | 4.13.7(2) | | | | NE | | The addition or deletion of backing gas, a change in backing gas composition, or a change in the specified flow rate range of the backing gas | QW-408.5 | | NE | NE | | |
| | | | | | | | For groove welds in P-No. 41 through P-No. 49 and all welds of P-No. 10I, P-No. 10J, P-No. 10K, P-No. 51 through P-No. 53, and P-No. 61 through P-No. 62 metals, the deletion of backing gas or a change in the nominal composition of the backing gas from an inert gas to a mixture including non-inert gas(es) | QW-408.9 | | E | E | | |

(continued)

**Table 5.7** (continued)

| AWS B2.1 | | | | | | | ASME IX | | | | | | |
|---|---|---|---|---|---|---|---|---|---|---|---|---|---|
| Variable | Ref. | SMAW | GMAW | FCAW | GTAW | SAW | Variable | Ref. | SMAW | GMAW | FCAW | GTAW | SAW |
| Addition or deletion of torch shielding gas | 4.14.7(1) | | E | E | E | | A separate procedure qualification is required for each of the following:<br>(a) a change from a single shielding gas to any other single shielding gas<br>(b) a change from a single shielding gas to a mixture of shielding gasses, and vice versa<br>(c) a change in the specified percentage composition of a shielding gas mixture<br>(d) the addition or omission of shielding gas<br>The AWS classification of SFA-5.32 may be used to specify the shielding gas composition | QW-408.2 | | E | E | | |
| A change in the specified nominal composition of shielding gas | 4.14.7(2) | | E | E | E | | | | | | | | |
| For M-51, M-52, M-61, and M-10I base metal, a change in the nominal composition or a decrease of 15% in the root shielding flow rate | 4.14.7(6) | | E | | E | | N/A | | | | | | |
| For M-21 through M-27, an increase of 50% or more, or a decrease of 20% or more in the shielding gas flow rate used for qualification | 4.14.7(7) | | E | | E | | N/A | | | | | | |

(continued)

**Table 5.7** (continued)

| AWS B2.1 | | | | | | | ASME IX | | | | | | |
|---|---|---|---|---|---|---|---|---|---|---|---|---|---|
| Variable | Ref. | SMAW | GMAW | FCAW | GTAW | SAW | Variable | Ref. | SMAW | GMAW | FCAW | GTAW | SAW |
| The addition, deletion, or a change in composition, or a decrease exceeding 15% in the flow rate of root shielding gas on single-sided M-4X groove joints and for all welds in M-51 through M-54, M-61 and M-62, M-10I, M-10J, and M-10K | 4.14.7(8) | | E | | E | | N/A | | | | | | |
| For M-10I, M-51 through M-54, and M-61 and M-62, the deletion of, or a change in composition of, or a decrease exceeding 10% in the trailing gas flow rate | 4.14.7(9) | | E | | E | | N/A | | | | | | |
| N/A | | | | | | | The addition or deletion of trailing gas and/or a change in its composition | QW-408.1 | | NE | NE | | |
| N/A | | | | | | | For P-No. 10I, P-No. 10J, P-No. 10K, P-No. 51 through P-No. 53, and P-No. 61 through P-No. 62 metals, the deletion of trailing gas, or a change in the nominal composition of the trailing gas from an inert gas to a mixture including non-inert gas(es), or a decrease of 10% or more in the trailing gas flow rate | QW-408.10 | | E | E | | |

**Table 5.8** Electrical characteristics variables for a groove or fillet welding procedure

| AWS B2.1 | | | | | | | ASME IX | | | | | | |
|---|---|---|---|---|---|---|---|---|---|---|---|---|---|
| Variable | Ref. | SMAW | GMAW | FCAW | GTAW | SAW | Variable | Ref. | SMAW | GMAW | FCAW | GTAW | SAW |
| Current (or wire feed speed), current type, and polarity | 4.13.8(1) | NE | NE | NE | NE | NE | A change from AC to DC, or vice versa; and in DC welding, a change from electrode negative (straight polarity) to electrode positive (reverse polarity), or vice versa. This variable does not apply to a WPS qualified for welding base metals that are assigned to P-No. 8, P-Nos. 21 through 26, and P-Nos. 41 through 49 | QW-409.4 | NE | NE | NE | NE | NE |
| Voltage range (except for manual welding) | 4.13.8(2) | | NE | NE | NE | NE | A change in the range of amperage, or except for SMAW, GTAW, or waveform controlled welding, a change in the range of voltage. A change in the range of electrode wire feed speed may be used as an alternative to amperage. See Nonmandatory Appendix H | QW-409.8 | NE | NE | NE | NE | NE |
| Specification, classification, and diameter of tungsten electrode | 4.13.8(4) | | | | NE | | A change in type or size of tungsten electrode | QW-409.12 | | | | NE | |

(continued)

**Table 5.8** (continued)

| AWS B2.1 | | | | | | | ASME IX | | | | | | |
|---|---|---|---|---|---|---|---|---|---|---|---|---|---|
| Variable | Ref. | SMAW | GMAW | FCAW | GTAW | SAW | Variable | Ref. | SMAW | GMAW | FCAW | GTAW | SAW |
| Transfer mode | 4.13.8(5) | | NE | NE | | | A change from globular, spray or pulsed spray transfer welding to short-circuiting transfer welding or vice versa | QW-409.2 | | E | E | | |
| A change to or from pulsed current | 4.13.8(6) | | NE | NE | NE | | The addition or deletion of pulsing current to dc power source | QW-409.3 | | | | NE | |
| An increase in heat input, or an increase in volume of weld metal deposited per unit length of weld, over that qualified | 4.14.8 | SE | SE | SE | SE | SE | An increase in heat input, or an increase in volume of weld metal deposited per unit length of weld, for each process recorded on the PQR | QW-409.1 | SE | SE | SE | SE | SE |

**Table 5.9** Other variables for a groove or fillet welding procedure

| AWS B2.1 | | | | | | | ASME IX | | | | | | |
|---|---|---|---|---|---|---|---|---|---|---|---|---|---|
| Variable | Ref. | SMAW | GMAW | FCAW | GTAW | SAW | Variable | Ref. | SMAW | GMAW | FCAW | GTAW | SAW |
| Welding process and whether manual, semiautomatic, mechanized, or automatic | 4.13.9(1) | NE | NE | NE | NE | NE | A change from manual or semiautomatic to machine or automatic welding and vice versa | QW-410.25 | NE | NE | NE | NE | NE |
| A change in welding process | 4.14.9(1) | E | E | E | E | E | A change from one welding process to another welding process is an essential variable and requires requalification | QW-401 | E | E | E | E | E |
| For mechanized or automatic, single or multiple electrode and spacing | 4.13.9(2) | | NE | NE | NE | NE | A change in the spacing of multiple electrodes for machine or automatic welding | QW-410.15 | | NE | NE | NE | NE |
| A change in the number of electrodes in the same weld pool | 4.14.9(2) | | SE | SE | SE | SE | A change from single electrode to multiple electrode, or vice versa, for machine or automatic welding only. This variable does not apply when a WPS is qualified with a PWHT above the upper transformation temperature or when an austenitic or P-No. 10H material is solution annealed after welding | QW-410.10 | | SE<br>NE | SE<br>NE | SE<br>NE | SE<br>NE |
| Single or multipass | 4.13.9(3) | NE | NE | NE | NE | NE | A change from multipass per side to single pass per side. This variable does not apply for any of the following conditions:<br>(a) WPS is qualified with a heat treatment above the upper transformation temperature<br>(b) WPS is for welding austenitic or P-10H material and is qualified with a solution heat treatment<br>(c) Base metals are assigned to P-No. 8, P-Nos. 21 through 26, and P-Nos. 41 through 49 | QW-410.9 | SE<br>NE | SE<br>NE | SE<br>NE | SE<br>NE | SE<br>NE |
| A change from multipass per side to single pass per side | 4.14.9(3) | SE | SE | SE | SE | SE | | | | | | | |
| Contact tube to work distance | 4.13.9(4) | | NE | NE | | NE | A change in the contact tube to work distance | QW-410.8 | | NE | NE | | NE |
| N/A | | | | | | | A change in the orifice, cup, or nozzle size | QW-410.3 | | NE | NE | NE | |
| Cleaning | 4.13.9(5) | NE | NE | NE | NE | NE | A change in the method of initial and interpass cleaning (brushing, grinding, etc.) | QW-410.5 | NE | NE | NE | NE | NE |
| Peening | 4.13.9(6) | NE | NE | NE | NE | NE | The addition or deletion of peening | QW-410.26 | NE | NE | NE | NE | NE |
| Stringer or weave bead | 4.13.9(11) | NE | NE | NE | NE | NE | For manual or semiautomatic welding, a change from the stringer bead technique to the weave bead technique, or vice versa | QW-410.1 | NE | NE | NE | NE | NE |

(continued)

**Table 5.9** (continued)

| AWS B2.1 | | | | | | | ASME IX | | | | | | |
|---|---|---|---|---|---|---|---|---|---|---|---|---|---|
| Variable | Ref. | SMAW | GMAW | FCAW | GTAW | SAW | Variable | Ref. | SMAW | GMAW | FCAW | GTAW | SAW |
| A change from a stringer to a weave bead, but not vice versa, for M-23, M-24, M-26, and M-27 heat-treatable materials | 4.14.9(12) | | E | | E | | N/A | | | | | | |
| A change from a stringer bead to a weave bead in vertical uphill welding | 4.14.9(13) | SE | SE | SE | SE | | N/A | | | | | | |
| Travel-speed range for mechanized or automatic welding and manual applications requiring heat input calculations | 4.13.9(12) | NE | NE | NE | NE | NE | (see QW-409.1) | | | | | | |
| A change exceeding ±20% in the oscillation variables for mechanized or automatic welding | 4.14.9(9) | | SE | SE | SE | SE | For the machine or automatic welding process, a change in width, frequency, or dwell time of oscillation technique | QW-410.7 | | NE | NE | NE | NE |
| N/A | | | | | | | A change in the method of back gouging | QW-410.6 | NE | NE | NE | NE | NE |
| N/A | | | | | | | For vessels or parts of vessels constructed with P-No. 11A and P-No. 11B base metals, weld grooves for thicknesses less than 5/8 in. (16 mm) shall be prepared by thermal processes when such processes are to be employed during fabrication. This groove preparation shall also include back gouging, back grooving, or removal of unsound weld metal by thermal processes when these processes are to be employed during fabrication | QW-410.64 | E | E | E | E | E |
| N/A | | | | | | | A change from closed chamber to out-of-chamber conventional torch welding in P-No. 51 through P-No. 53 metals, but not vice versa | QW-410.11 | | | | E | |

## 5.2 Examination and Testing Requirements

The following sections separately outline the examination and testing requirements for groove weld and fillet weld procedure qualification.

### 5.2.1 Groove Welds Test Weldments

The testing methods and the number of tests required for a groove weld procedure qualification test for both AWS B2.1 (ref. 4.7 and Table 4.1) and ASME IX (ref. QW-202 and Table QW-451.1) are summarized in Table 5.10.

Special test weldments shall be examined and tested as specified by the referencing document. When a test in Table 5.10 is specified by the referencing document, the acceptance criteria shall be as required in AWS B2.1 or ASME IX. The criteria for all other tests shall be as specified in the referencing document.

Depending on the application of the welding procedure, material type(s), or additional requirements or standards invoked by the referencing document, additional testing may be specified. Additional testing may include but is not limited to:

- Hardness Testing
- Ferrite Number Testing
- Macro or Micro Examination.

*Visual Examination for AWS B2.1*

Unless otherwise specified in the referencing document(s), the visual examination acceptance criteria shall be as follows.

For base metal thickness $\geq$ 1/16 in. [1.5 mm] (ref. AWS B2.1 4.7.3):

- There shall be no evidence of cracks, incomplete fusion, or incomplete joint penetration.
- The depth of undercut shall not exceed the lesser of 10% of the base metal thickness or 1/32 in. [1 mm].

**Table 5.10** Test methods required for groove weld procedure qualification

| | AWS B2.1 | ASME IX |
|---|---|---|
| Visual examination | Yes | Not required |
| Tension testing | Yes (qty. 2) | Yes (qty. 2) |
| Bend testing | Yes (qty. 4) | Yes (qty. 4) |
| Toughness testing | If specified | If specified |

- Porosity shall not exceed the limitations of the referencing document.

For base metal thickness<1/16 in. [1.5 mm] (ref. AWS B2.1 4.12.2):

- No incomplete joint penetration or incomplete fusion.
- Not more than one visible pore or inclusion exceeding 0.25 of the base metal thickness, shall be permitted in any 1 in. [25 mm] of weld.
- Visible pores shall not extend through the weld thickness.
- The weld reinforcement shall not exceed 1/8 in. [3 mm].
- Undercut shall not exceed 0.15 times the thickness of the test weldment base metal.

*Visual Examination for ASME IX*
There is no requirement for visual examination.

*Tension Testing for AWS B2.1*
The mechanical testing specimen blanks shall be removed from the locations shown in AWS B2.1 Fig. 4.2 for pipe, Fig. 4.3 for box tube, and Figs. 4.4 or 4.5 for plate. The preparation and dimensions of specimen blanks for tension testing are provided in AWS B2.1 Annex A Figure A.3A for rectangular reduced section tension specimens, Figure A.3B for round reduced section tension specimens, Figure A.3C for alternate tension specimens in 3 in. [76 mm] O.D. or less pipe, and Figure A.3D for alternate tension specimens in 2 in. [51 mm] O.D. or less pipe.

For thicknesses up to and including 1 in. [25 mm], a full thickness specimen shall be used for each required tension test. For thicknesses over 1 in. [25 mm], single or multiple specimens may be used provided that: (1) Collectively, multiple specimens, representing the full thickness of the weld at one location, shall comprise a set. (2) The entire thickness shall be mechanically cut into approximately equal strips. For specimens that are not turned, specimen thicknesses shall be the maximum size that can be tested in available equipment (ref. AWS B2.1 Annex A3).

The acceptance criteria states that each tensile test specimen shall have a tensile strength not less than the following (ref. AWS B2.1 4.7.5):

- The specified minimum tensile strength of the base metal, or of the weaker of the two base metals if metals of different minimum tensile strength are used; or
- The specified minimum tensile strength of the weld metal when the referencing document provides for the use of weld metal having a lower tensile strength than the base metal; or
- If the specimen breaks in the base metal outside of the weld or weld interface, the test shall be accepted, provided the strength is not more than 5% below the specified minimum tensile strength of the base metal; or

- If the base metal has no specified minimum tensile strength, then failure in the base metal shall be acceptable.

*Tension Testing for ASME IX*

The mechanical testing specimen blanks shall be removed from locations shown in ASME IX Figures QW-463.1(d), (e), and (f) for pipe, and Figures QW-463.1(a), (b), and (c) for plate. The preparation and dimensions of specimen blanks for tension testing are provided in ASME IX Figure QW-462.1(a) for reduced section tension specimens in plate, Figure QW-462.1(b) for reduced section tension specimens in greater than 3 in. [76 mm] O.D. pipe, Figure QW-462.1(c) for alternate reduced section tension specimens in 3 in. [76 mm] O.D. or less pipe, Figure QW-462.1(d) for round reduced section tension specimens, and Figure QW-462.1(e) for full section tension specimens in 3 in. [76 mm] O.D. or less pipe.

The acceptance criteria in ASME IX QW-153 states that each tensile test specimen shall have a tensile strength that is not less than:

- the minimum specified tensile strength of the base metal; or
- the minimum specified tensile strength of the weaker of the two, if base metals of different minimum tensile strengths are used; or
- the minimum specified tensile strength of the weld metal when the applicable Section provides for the use of weld metal having lower room temperature strength than the base metal;
- if the specimen breaks in the base metal outside of the weld or weld interface, the test shall be accepted as meeting the requirements, provided the strength is not more than 5% below the minimum specified tensile strength of the base metal.
- the specified minimum tensile strength is for full thickness specimens including cladding for Aluminum Alclad materials (P-No. 21 through P-No. 23) less than 1/2 in. [13 mm]. For Aluminum Alclad materials 1/2 in. [13 mm] and greater, the specified minimum tensile strength is for both full thickness specimens that include cladding and specimens taken from the core.

Minimum values for procedure qualification are provided under the column heading "Minimum Specified Tensile, ksi" of ASME IX Table QW/QB-422.

*Bend Testing for AWS B2.1*

The mechanical testing specimen blanks shall be removed from the locations shown in AWS B2.1 Fig. 4.2 for pipe, Fig. 4.3 for box tube, and Figs. 4.4 or 4.5 for plate. The preparation and dimensions of specimen blanks for bend testing are provided in AWS B2.1 Annex A Figure A.2C for transverse side bend specimens, Figure A.2A for transverse face and root bend specimens, and Figure A.2B for longitudinal face and root bend specimens. The cut surfaces of AWS B2.1 Figures A.2A and A.2B are designated the specimen sides and the other two surfaces are designated the face and root surfaces. Dimensions for subsize transverse face and root bends, and for nonstandard bend specimens are provided in AWS B2.1 Annex A A2.3 and A2.4 respectively.

Weld reinforcement and backing of face- and root- bend specimens shall be removed flush with the specimen surface. Cut surfaces shall be parallel, may be thermally cut, and shall be machined or ground a minimum of 1/8 in. [3 mm] on thermally cut edges, except that M-1 metals may be bent "as-cut" if no objectionable surface roughness exists.

Bend specimens shall be bent in one of the guided bend test fixtures shown in AWS B2.1 Annex A Figure A.5A bottom ejecting guided bend fixture, Figure A.5B bottom type guided bend fixture, or Figure A.5C wrap around guided bend fixture. For transverse specimens, the weld metal and heat-affected zone shall be completely within the bent portion of the specimen after bending.

The acceptance criteria for bend testing in AWS B2.1 4.7.6 states:

- There shall be no open discontinuity exceeding 1/8 in. [3 mm], measured in any direction on the convex surface of the specimen after bending.
- Cracks occurring on the corners of the specimen during bending shall not be considered, unless there is definite evidence that they result from weld discontinuities.

*Bend Testing for ASME IX*

The mechanical testing specimen blanks shall be removed from locations shown in ASME IX Figures QW-463.1(d), (e), and (f) for pipe, and Figures QW-463.1(a), (b), and (c) for plate. The preparation and dimensions of specimen blanks for bend testing are provided in ASME IX QW-161.1 and Figure QW-462.2 for transverse side bend specimens, QW-161.2, QW-161.3 and Figure QW-462.3(a) for transverse face and root bend specimens, and QW-161.6, QW161.7 and Figure QW-462.3(b) for longitudinal face and root bend specimens. Dimensions for subsize transverse face and root bend specimens are outlined in QW-161.4 and Figure QW-462.3(a) general note (b). Weld reinforcements and backing strip or backing ring, if any, shall be removed essentially flush with the undisturbed surface of the base material.

Bend specimens shall be bent in one of the guided bend test fixtures shown in ASME IX Figure QW-466.1 test jig dimensions, Figure QW-466.2 guided-bend roller jig, or Figure QW-466.3 guided-bend wrap around jig. The weld and heat-affected zone of a transverse weld bend specimen shall be completely within the bent portion of the specimen after testing.

The acceptance criteria for bend testing in ASME IX QW-163 states:

- The guided-bend specimens shall have no open discontinuity in the weld or heat-affected zone exceeding 1/8 in. [3 mm], measured in any direction on the convex surface of the specimen after bending.

- Open discontinuities occurring on the corners of the specimen during testing shall not be considered unless there is definite evidence that they result from lack of fusion, slag inclusions, or other internal discontinuities.

*Toughness Testing for AWS B2.1*

If the referencing document requires fracture toughness testing, then procedure qualification shall be made for each combination M-Number and Group Number to be joined. A procedure qualification shall be made for each M-Number and Group Number combination of base metals, even though procedure qualification tests have been made for each of the two base metals welded to itself (ref. AWS B2.1 4.3.14).

For fracture toughness testing, the type of test, number of specimens, and acceptance criteria shall be in accordance with the referencing document. The procedures and apparatus shall conform to the requirements of AWS B4.0, *Standard Methods for Mechanical Testing of Welds* (ref. AWS B2.1 4.7.7).

When fracture toughness is a requirement and a qualified procedure exists which satisfies all requirements except fracture toughness, it is necessary only to prepare an additional test weldment with sufficient material to provide the required fracture toughness specimens. The test plate shall be welded using that procedure, plus those variables applicable to fracture toughness. A new or revised PQR shall be prepared and the WPS shall then be revised or a new WPS issued to accommodate the qualification variables for fracture toughness applications (ref. AWS B2.1 4.3.17).

*Toughness Testing for ASME IX*

Toughness tests shall be made when required by referencing codes. Test procedures and apparatus shall conform to the requirements of the referencing code. When not specified by the referencing code, the test procedures and apparatus shall conform to the requirements of ASME BPVC Sec. II Part A SA-370 (ref. ASME IX QW-171.1). The toughness test specimen removal and preparation requirements shall be as given in the Section requiring such tests (ref. ASME IX QW-171.3). The acceptance criteria shall be in accordance with that Section specifying toughness testing requirements (ref. ASME IX QW-171.2).

### 5.2.2 Fillet Weld Test Weldments

The testing methods and the number of tests required for a fillet weld procedure qualification test for both AWS B2.1 (ref. 4.7 and Table 4.1) and ASME IX (ref. QW-180 and Table QW-451.3) are summarized in Table 5.11.

AWS B2.1 allows utilization of a groove-weld test coupon to qualify fillet welds, but not vice versa (ref. AWS B2.1 4.8.1). ASME IX requires pressure retaining fillet welds be qualified on groove-weld test coupons but allows non-pressure retaining fillet welds to be qualified by a fillet weld test coupon (ref. QW-202.2(c)). For fillet welds qualified on groove-weld test coupons, the examination and testing requirements for groove weld apply as outlined in the previous Sect. 5.2.1. For fillet welds qualified on fillet weld test coupons, the following sections outline the examination and testing requirements.

Special test weldments shall be examined and tested as specified by the referencing document. When a test in Table 5.11 is specified by the referencing document, the acceptance criteria shall be as required in AWS B2.1 or ASME IX. The criteria for all other tests shall be as specified in the referencing document.

Depending on the application of the welding procedure, material type(s), or additional requirements or standards invoked by the referencing document, additional testing may be specified. Additional testing may include but is not limited to:

- Fatigue Testing
- Hardness Testing.

*Visual Examination for AWS B2.1*

The acceptance criteria for visual examination stated in AWS B2.1 4.8.3 is as follows:

- There shall be no cracks or incomplete fusion.
- Undercut depth shall not exceed the lesser of 10% of the base metal thickness or 1/32 in. [1 mm].
- Profile and porosity limitations shall be in accordance with the referencing document.

*Visual Examination for ASME IX*

There is no requirement for visual examination.

**Table 5.11** Test methods required for fillet weld procedure qualification

| | AWS B2.1 | ASME IX |
|---|---|---|
| Visual examination | Yes | Not required |
| Macro examination | Yes (qty. 2) | Yes (qty. 4 or 5) |
| Bend-break testing | Yes | Not required |
| Shear testing | Alternate to bend-break | Not required |

*Macro Examination for AWS B2.1*

The specimen blanks for macro examination shall be removed from the locations shown in AWS B2.1 Figs. 4.6 or 4.7. AWS B2.1 4.8.4 states there shall be no cracks, incomplete joint penetration, or incomplete fusion in the macro cross section.

*Macro Examination for ASME IX*

The specimen blanks for macro examination shall be removed from locations shown in ASME IX Figure QW-462.4(a) for fillet welds in plates and Figure QW-462.4(d) for fillet welds in pipe. Five test specimens are required for fillet weld test coupons in plate and four test specimens are required for fillet weld test coupons in pipe (ref. ASME IX QW-183). The examination of the cross sections shall include only one side of the test specimen at the area where the plate or pipe is divided into sections i.e., adjacent faces at the cut shall not be used. Visual examination of the cross sections of the weld metal and heat-affected zone shall show complete fusion and freedom from cracks. There shall be not more than 1/8 in. [3 mm] difference in the length of the legs of the fillet.

*Bend-Break Testing for AWS B2.1*

If both single and multiple pass welds are to be qualified, one procedure qualification specimen shall be welded with the maximum size single pass to be used, and a second shall be welded with the minimum size multiple pass to be used. The specimen blanks for bend-break testing shall be removed from the locations shown in AWS B2.1 Fig. 4.7. The test specimens shall be bent with the weld root in tension until the specimen either fractures or until it is bent flat upon itself. The specimen shall be accepted if it does not fracture or if the fillet fractures, the fractured surface shall exhibit no cracks or incomplete root fusion and no inclusion or porosity in the fracture surface exceeding 3/32 in. [2 mm] in its greatest dimension. The sum of the greatest dimension of all inclusions and porosity shall not exceed 3/8 in. [10 mm] in the specimen length (ref. AWS B2.1 4.8.5).

*Bend-Break Testing for ASME IX*

There is no requirement for bend-break testing.

*Shear Testing for AWS B2.1*

The specimen blanks for shear testing shall be removed from the locations shown in AWS B2.1 Fig. 4.6. Unless otherwise stated in the referencing document, the fillet shear strength shall be not less than 60% of the lower of the minimum specified tensile strength of the base metal or weld metal. If neither value is available, two specimen blanks of the base material shall be tension tested. The lowest value determined from these tests shall be the specified minimum tensile strength for qualification purposes (ref. AWS B2.1 4.8.6).

*Shear Testing for ASME IX*

There is no requirement for shear testing.

## 5.3 Qualification Limits

In addition to the restrictions of specific variables for the different welding processes as outlined in Table 5.2 through Table 5.9, additional limitations for Process Control, Position, Base Metal Thickness, and Weld Deposit Thickness are discussed below.

### 5.3.1 Process Control

For groove and fillet weld procedure qualifications, both AWS B2.1 and ASME IX allows qualification by any form of process control, whether manual, semiautomatic, mechanized/machine, or automatic, to qualify the welding procedure for all forms of process control (ref. AWS B2.1 4.13.9(1) and ASME IX QW-410.25). The qualified welding procedure may however restrict a change from one form of process control to another.

### 5.3.2 Position

For groove and fillet weld procedure qualifications, both AWS B2.1 and ASME IX allows qualification in any position to qualify the welding procedure for all positions (ref. AWS B2.1 4.3.10 and ASME IX QW-203). The standard welding positions (1G Flat, 2G Vertical, 3G Horizontal, 4G Overhead) for groove welds in plate or pipe are defined in AWS B2.1 Figure A.1A and ASME IX Figures QW-461.3 and QW-461.4. The standard welding positions (1F Flat, 2F Vertical, 3F Horizontal, 4F Overhead) for fillet welds in plate or pipe are defined in AWS B2.1 Figure A.1C and ASME IX Figures QW-461.5 and QW-461.6.

Both AWS B2.1 and ASME IX define a change in position as an angular deviation of $>\pm 15$ degrees from the specified horizontal and vertical planes, and an angular deviation of $>\pm 5$ degrees from the specified inclined plane are permitted during welding. This is depicted for groove welds in AWS B2.1 Figure A.1B and ASME IX QW-461.1, while depictions for fillet welds are in AWS B2.1 Figure A.1D and ASME IX QW-461.2.

Additional restrictions on vertical progression for the 3G and 3F vertical position are outlined in Table 5.4 for AWS B2.1 and ASME IX.

### 5.3.3 Base Metal Thickness

The base metal thickness range qualified for groove and fillet welding procedures (WPS) are governed by the qualification test weldment base metal thickness.

*Groove Weld Base Metal Thickness for AWS B2.1*

The qualified groove weld base metal thickness ranges for AWS B2.1 are shown in Table 5.12. It should be noted that AWS B2.1 does not differentiate separate qualification ranges for test weldments qualified by transverse or longitudinal bend tests.

Additional limitations and deviations from the base metal thickness qualification ranges are outlined in the footnotes of AWS B2.1 Table 4.3.

*Groove Weld Base Metal Thickness for ASME IX*

The qualified groove base metal weld thickness ranges for ASME IX are shown in Tables 5.13 and 5.14. It should be noted that ASME IX differentiates separate qualification ranges for test weldments qualified by transverse or longitudinal bend tests.

Additional limitations and deviations from the base metal thickness qualification ranges are outlined in the footnotes of ASME IX Tables QW-451.1 and QW-451.2.

**Table 5.12** Qualified groove weld base metal thickness for all bend tests (ref. AWS B2.1 Table 4.3)

| Test weldment thickness (*T*) in. [mm] | Base metal thickness qualified | |
|---|---|---|
| | Minimum in. [mm] | Maximum in. [mm] |
| <1/16 [1.5] | 1/2 *T* | 2 *T* |
| 1/16 to 3/8 [1.5 to 10] | 1/16 [1.5] | 2 *T* |
| >3/8 to < 3/4 [>10 to < 19] | 3/16 [5] | 2 *T* |
| 3/4 to < 1 1/2 [19 to < 38] | 3/16 [5] | 2 *T* |
| 1 1/2 to < 6 [38 to < 152] | 3/16 [5] | 8 [203] |
| ≥6 [152] | 1 [25] | 1.33 *T* |

**Table 5.13** Qualified groove weld base metal thickness for transverse bend tests (ref. ASME IX Table QW-451.1)

| Test weldment thickness (*T*) in. [mm] | Base metal thickness qualified | |
|---|---|---|
| | Minimum in. [mm] | Maximum in. [mm] |
| <1/16 [1.5] | *T* | 2 *T* |
| 1/16 to 3/8 [1.5 to 10] | 1/16 [1.5] | 2 *T* |
| >3/8 to < 3/4 [>10 to < 19] | 3/16 [5] | 2 *T* |
| 3/4 to < 1 1/2 [19 to < 38] | 3/16 [5] | 2 *T* |
| 1 1/2 to ≤ 6 [38 to ≤ 152] | 3/16 [5] | 8 [203] |
| >6 [150] | 3/16 [5] | 1.33 *T* |

**Table 5.14** Qualified groove weld base metal thickness for longitudinal bend tests (ref. ASME IX Table QW-451.2)

| Test weldment thickness ($T$) in. [mm] | Base metal thickness qualified | |
|---|---|---|
| | Minimum in. [mm] | Maximum in. [mm] |
| <1/16 [1.5] | $T$ | 2 $T$ |
| 1/16 to 3/8 [1.5 to 10] | 1/16 [1.5] | 2 $T$ |
| >3/8 [10] | 3/16 [5] | 2 $T$ |

**Table 5.15** Qualified fillet base metal weld thickness for groove weld test weldments (ref. AWS B2.1 Table 4.3 note h)

| Test weldment thickness ($T$) in. [mm] | Base metal thickness qualified | |
|---|---|---|
| | Minimum in. [mm] | Maximum in. [mm] |
| <1/16 [1.5] | 1/2 $T$ | 2 $T$ |
| 1/16 to 3/8 [1.5 to 10] | 1/16 [1.5] | 2 $T$ |
| >3/8 | Unlimited | |

**Table 5.16** Qualified fillet weld base metal thickness for fillet weld test weldments (ref. AWS B2.1 Table 4.2)

| Fillet test weldment | Base metal thickness qualified |
|---|---|
| Single pass | Unlimited |
| Multiple pass | Unlimited |

*Fillet Weld Base Metal Thickness for AWS B2.1*

For fillet welds qualified by a groove weld test weldment, the base metal thickness qualification range is shown in Table 5.15. For fillet welds qualified by a fillet weld test weldment, the base metal thickness qualification range is shown in Table 5.16.

Additional limitations and deviations from the base metal thickness qualification ranges are outlined in the footnotes of AWS B2.1 Tables 4.2 and 4.3.

*Fillet Weld Base Metal Thickness for ASME IX*

For fillet welds qualified by either by a groove weld test weldment or fillet weld test weldment, the base metal thickness qualification range is unlimited (ref. ASME IX Tables QW-451.3 and QW-451.4).

### 5.3.4 Weld Deposit Thickness

The weld deposit thickness range qualified for groove and fillet welding procedures (WPS) are also governed by the qualification test weldment base metal thickness.

**Table 5.17** Qualified groove weld deposit thickness for all bend tests (ref. AWS B2.1 Table 4.3)

| Test weldment thickness ($T$) in. [mm] | Weld metal deposit thickness qualified |
|---|---|
| | Maximum in. [mm] |
| <1/16 [1.5] | $2t$ |
| 1/16 to 3/8 [1.5 to 10] | $2t$ |
| >3/8 to < 3/4 [>10 to < 19] | $2t$ |
| 3/4 to < 1 1/2 [19 to < 38] | $2t$ when $t < 3/4$ [19]<br>2 $T$ when $t \geq 3/4$ [19] |
| 1 1/2 to < 6 [38 to < 152] | $2t$ when $t < 3/4$ [19]<br>8 [203] when $t \geq 3/4$ [19] |
| ≥6 [152] | $2t$ when $t < 3/4$ [19]<br>8 [203] when 3/4 [19] $\leq t <$ 6 [152]<br>$1.33t$ when $t \geq 6$ [152] |

When multiple welding processes are utilized within the same test weldment, the maximum weld metal deposit thickness for each process is determined by the amount of weld metal deposited by each process in the test weldment (ref. AWS B2.1 Table 4.3 note b(2) and ASME IX interpretation BC-80-628).

*Groove Weld Deposit Thickness for AWS B2.1*

The qualified groove weld deposit thickness ranges for AWS B2.1 are shown in Table 5.17. It should be noted that AWS B2.1 does not differentiate separate qualification ranges for test weldments qualified by transverse or longitudinal bend tests.

Additional limitations and deviations from the weld metal deposit thickness qualification ranges are outlined in the footnotes of AWS B2.1 Table 4.3.

*Groove Weld Deposit Thickness for ASME IX*

The qualified groove weld deposit thickness ranges for ASME IX are shown in Tables 5.18 and 5.19. It should be noted that ASME IX differentiates separate qualification ranges for test weldments qualified by transverse or longitudinal bend tests.

Additional limitations and deviations from the weld metal thickness qualification ranges are outlined in the footnotes of ASME IX Tables QW-451.1 and QW-451.2.

*Fillet Weld Deposit Thickness for AWS B2.1*

For fillet welds qualified by a groove weld test weldment, the weld deposit thickness qualification range is shown in Table 5.20. For fillet welds qualified by a fillet weld test weldment, the weld deposit thickness qualification range is shown in Table 5.21.

Additional limitations and deviations from the weld metal thickness qualification ranges are outlined in the footnotes of AWS B2.1 Tables 4.2 and 4.3.

**Table 5.18** Qualified groove weld deposit thickness for transverse bend tests (ref. ASME IX Table QW-451.1)

| Test weldment thickness ($T$) in. [mm] | Weld metal deposit thickness qualified |
|---|---|
| | Maximum in. [mm] |
| <1/16 [1.5] | $2t$ |
| 1/16 to 3/8 [1.5 to 10] | $2t$ |
| >3/8 to < 3/4 [>10 to < 19] | $2t$ |
| 3/4 to < 1 1/2 [19 to < 38] | $2t$ when $t < 3/4$ [19]<br>2 $T$ when $t \geq 3/4$ [19] |
| 1 1/2 to ≤ 6 [38 to ≤ 152] | $2t$ when $t < 3/4$ [19]<br>8 [203] when $t \geq 3/4$ [19] |
| >6 [150] | $2t$ when $t < 3/4$ [19]<br>1.33 $T$ when $t \geq 3/4$ [19] |

**Table 5.19** Qualified groove weld deposit thickness for longitudinal bend tests (ref. ASME IX Table QW-451.2)

| Test weldment thickness ($T$) in.[mm] | Weld metal deposit thickness qualified |
|---|---|
| | Maximum in.[mm] |
| <1/16 [1.5] | $2t$ |
| 1/16 to 3/8 [1.5 to 10] | $2t$ |
| >3/8 [10] | $2t$ |

**Table 5.20** Qualified fillet weld deposit thickness for groove weld test weldments (ref. AWS B2.1 Table 4.3 note g)

| Test weldment thickness (T) in. [mm] | Weld metal deposit thickness qualified |
|---|---|
| | Maximum in. [mm] |
| < = 3/8 [10] | $2t$ |
| > 3/8 [10] | Unlimited |

**Table 5.21** Qualified fillet weld deposit thickness for fillet weld test weldments (ref. AWS B2.1 Table 4.2)

| Fillet test weldment | Weld metal deposit thickness qualified |
|---|---|
| Single pass | Maximum welded single-pass fillet size and smaller |
| Multiple pass | 1/2 of that welded during qualification to unlimited |

*Fillet Weld Deposit Thickness for ASME IX*

For fillet welds qualified by either by a groove weld test weldment or a fillet weld test weldment, the qualified weld deposit thickness range is unlimited (ref. ASME IX Tables QW-451.3 and QW-451.4).

## 5.4 Summary

This chapter discussed the specific requirements for the qualification of a groove or fillet welding procedures in accordance with the rules of AWS B2.1 and ASME IX. The required welding variables to be documented in the Procedure Qualification Record (PQR) are discussed for each welding process. The required examination and test methods along with their acceptance criteria are also outlined. Finally, the required welding variables to be addressed in the Welding Procedure Specification (WPS) along with their qualified limits and ranges are discussed.

# 6 Procedure Qualification for Corrosion Resistant Overlay

The concepts of developing a qualified welding procedure and producing the required welding documentation was introduced in Chap. 4. This chapter discusses the specific requirements for the qualification of a corrosion resistant overlay weld procedure. The required documented welding variables for the Procedure Qualification Record (PQR) and their qualification limits for the Welding Procedure Specification (WPS) are discussed, along with the required examination and test methods and their acceptance criteria are outlined.

## 6.1 Corrosion Resistant Overlay Variables

The following tables provide the required welding variables for fillet and groove welding procedures qualified under AWS B2.1 (ref. Tables 4.1.3 and 4.1.4) and ASME IX (ref. Tables QW-253.1, QW-254.1, QW-255.1, & QW-256.1 and Article IV). For convenience, the variables are grouped as discussed in Section 4.1.9. The essential (E), supplementary essential (SE), and non-essential (NE) variables are compared for the five most common arc welding procedures (SMAW, GMAW, FCAW, GTAW, SAW). Variables marked N/A are not addressed for a given standard. The tables should be regarded as illustrative only since the various standards are revised periodically.

© The Author(s), under exclusive license to Springer Nature Switzerland AG 2025

D. Barborak, *Arc Welding Qualification Standards*, Synthesis Lectures on Welding Engineering, https://doi.org/10.1007/978-3-031-64646-1_6

### 6.1.1 Joint Design

See Table 6.1.

### 6.1.2 Base Metal

See Table 6.2.

### 6.1.3 Filler Metals

See Table 6.3.

### 6.1.4 Position

See Table 6.4.

### 6.1.5 Preheat and Interpass Temperature

See Table 6.5.

### 6.1.6 Postweld Heat Treatment

See Table 6.6.

### 6.1.7 Shielding Gas

See Table 6.7.

### 6.1.8 Electrical Characteristics

See Table 6.8.

### 6.1.9 Other Variables

See Table 6.9.

**Table 6.1** Joint design variables for a corrosion resistant overlay procedure

| AWS B2.1 | | | | | | | ASME IX | | | | | | |
|---|---|---|---|---|---|---|---|---|---|---|---|---|---|
| Variable | Ref. | SMAW | GMAW | FCAW | GTAW | SAW | Variable | Ref. | SMAW | GMAW | FCAW | GTAW | SAW |
| A decrease in thickness or a change in the nominal chemical composition of surfacing or buttering beyond that qualified | 4.14.3(9) | E | E | E | E | E | A decrease in the distance between the approximate weld interface and the final surface of the production corrosion-resistant or hard-facing weld metal overlay below the minimum thickness qualified as shown in Figures QW-462.5(a) through QW-462.5(e). There is no limit on the maximum thickness for corrosion resistant or hard-facing weld metal overlay that may be used in production | QW-402.16 | E | E | E | E | E |

**Table 6.2** Base metal variables for a corrosion resistant overlay procedure

| AWS B2.1 | | | | | | | ASME IX | | | | | | |
|---|---|---|---|---|---|---|---|---|---|---|---|---|---|
| Variable | Ref. | SMAW | GMAW | FCAW | GTAW | SAW | Variable | Ref. | SMAW | GMAW | FCAW | GTAW | SAW |
| M-number and group number | 4.13.2(1) | NE | NE | NE | NE | NE | If the chemical composition of the weld metal overlay is specified in the WPS, a change in the P-number listed in Table QW/QB-422 *to another* P-number or unlisted base metal, or a change in group number for P-No. 10 or P-No. 11 base metals<br>If the chemical composition of the weld metal overlay is not specified in the WPS, qualification on P-No. 5A or any lower P-number base metal also qualifies for weld metal overlay on all lower P-number base metals | QW-403.20 | E | E | E | E | E |
| Cladding and hardfacing require separate qualification for each base metal M-number, and filler metal combination | 4.3.17 | E | E | E | E | E | | | | | | | |
| A change from one M-number base metal to another M-number base metal or to a combination of M-number base metals except as permitted in 4.3.10 | 4.14.2(2) | E | E | E | E | E | | | | | | | |
| A change from one unlisted base metal to a different unlisted base metal or to a listed base metal or vice versa | 4.14.2(3) | E | E | E | E | E | | | | | | | |
| A change from one M-5 group (A, B, etc.) to any other. A change from M-9A to M-9B, but not vice versa. A change from M-10 to M-11 group (A, B, etc.) to any other group | 4.14.2(5) | E | E | E | E | E | | | | | | | |

(continued)

**Table 6.2** (continued)

| AWS B2.1 | | | | | | | ASME IX | | | | | | |
|---|---|---|---|---|---|---|---|---|---|---|---|---|---|
| Variable | Ref. | SMAW | GMAW | FCAW | GTAW | SAW | Variable | Ref. | SMAW | GMAW | FCAW | GTAW | SAW |
| Thickness range qualified | 4.13.2(2) | NE | NE | NE | NE | NE | A change in base metal thickness beyond the range qualified in Table QW-453 | QW-403.23 | E | E | E | E | E |
| A change in base metal thickness beyond the range permitted in 4.5 | 4.14.2(1) | E | E | E | E | E | | | | | | | |
| Limitations on the thickness ranges qualified by procedure qualification tests are given in the following tables:<br>Table 4.4—thickness limitations for cladding and hardfacing for procedure qualification | 4.5.1 | E | E | E | E | E | | | | | | | |
| The coating description or type, if present | 4.13.2(4) | NE | NE | NE | NE | NE | N/A | | | | | | |
| A change from an uncoated metal to a metal coated (such as painted or galvanized) metal unless the coating is removed from the weld area prior to welding, but not vice versa, as permitted in 4.3.8 | 4.14.2(6) | E | E | E | E | E | | | | | | | |

**Table 6.3** Filler metals variables for a corrosion resistant overlay procedure

| AWS B2.1 | | | | | | | ASME IX | | | | | | |
|---|---|---|---|---|---|---|---|---|---|---|---|---|---|
| Variable | Ref | SMAW | GMAW | FCAW | GTAW | SAW | Variable | Ref | SMAW | GMAW | FCAW | GTAW | SAW |
| Specification, classification, F- and A-number, or if not classified the nominal composition | 4.13.1(1) | NE | NE | NE | NE | NE | A change in the composition of the deposited weld metal from one A-number in Table QW-442 to any other A-number, or to an analysis not listed in the table. A change in the UNS number for each AWS classification of A-No. 8 or A-No. 9 analysis of Table QW-442, or each nonferrous alloy in Table QW-432, shall require separate WPS qualification. A-numbers may be determined in accordance with QW-404.5 | QW-404.37 | E | E | E | E | E |
| A change from one F-number to any other F-number or to any filler metal not listed in Annex B | 4.14.3(1) | E | E | E | E | E | | | | | | | |
| For ferrous materials, a change from one A-number to any other A-number or to a filler metal analysis not listed in Annex B (the PQR and WPS shall state the nominal chemical composition, the AWS specification and classification, or the manufacturer's designation for filler metals which do not fall in an A-number group). Qualification with A-1 shall qualify for A-2 and vice versa | 4.14.3(2) | E | E | E | E | E | | | | | | | |

(continued)

**Table 6.3** (continued)

| AWS B2.1 | | | | | | | ASME IX | | | | | | |
|---|---|---|---|---|---|---|---|---|---|---|---|---|---|
| Variable | Ref | SMAW | GMAW | FCAW | GTAW | SAW | Variable | Ref | SMAW | GMAW | FCAW | GTAW | SAW |
| For surfacing, a change in the chemical composition of the weld metal (A-number or alloy type). Each layer shall be considered independent of other layers | 4.14.3(3) | E | E | E | E | E | | | | | | | |
| A decrease in thickness or a change in the nominal chemical composition of surfacing or buttering beyond that qualified | 4.14.3(9) | E | E | E | E | E | | | | | | | |
| Flux-electrode classification | 4.13.3(4) | | | | | NE | For submerged-arc welding and electroslag welding, a change in the nominal composition or type of flux used. Requalification is not required for a change in flux particle size | QW-404.39 | | | | | E |
| If the weld metal alloy content is largely dependent upon the composition of the flux, any change in the welding procedure which would result in the important weld metal alloying elements being outside the specified chemical composition range of the WPS | 4.14.3(6) | | | | | E | | | | | | | |

(continued)

**Table 6.3** (continued)

| AWS B2.1 | | | | | | | ASME IX | | | | | | |
|---|---|---|---|---|---|---|---|---|---|---|---|---|---|
| Variable | Ref | SMAW | GMAW | FCAW | GTAW | SAW | Variable | Ref | SMAW | GMAW | FCAW | GTAW | SAW |
| A change from one AWS flux-electrode classification listed to any other electrode flux-electrode classification, or to an unlisted electrode-flux classification. A variation of 0.5% of the molybdenum content of the weld metal does not require requalification | 4.14.3(16) | | | | | E | | | | | | | |
| Weld metal thickness by process and filler metal classification | 4.13.3(2) | NE | NE | NE | NE | NE | N/A | | | | | | |
| A change in the weld metal thickness beyond that permitted in 4.5 | 4.14.3(17) | E | E | E | E | E | | | | | | | |
| Filler metal size or diameter<br>A change of filler metal/ electrode nominal size/ shape in the first layer | 4.14.3(3)<br>4.14.3(10) | NE<br>E | NE<br>E | NE<br>E | | NEE | A change in the nominal electrode diameter used for the first layer of deposit | QW-404.38 | NE | | | | |
| | | | | | | | A change in the nominal size of the electrode or electrodes specified in the WPS | QW-404.6 | | NE | NE | | NE |
| | | | | | | | A change in the size of the filler metal | QW-404.3 | | | | NE | |
| A change from wire to strip electrodes and vice versa | 4.14.3(15) | | | | | E | An increase in the nominal thickness or width of the electrode for strip filler metals used with the SAW and ESW processes for corrosion-resistant and hard-facing weld metal overlay | QW-404.57 | | | | | E |

(continued)

**Table 6.3** (continued)

| AWS B2.1 | | | | | | | ASME IX | | | | | | |
|---|---|---|---|---|---|---|---|---|---|---|---|---|---|
| Variable | Ref | SMAW | GMAW | FCAW | GTAW | SAW | Variable | Ref | SMAW | GMAW | FCAW | GTAW | SAW |
| N/A | | | | | | | A change from one of the following filler metal product forms to another:<br>(a) bare (solid or metal cored)<br>(b) flux cored<br>(c) flux coated (solid or metal cored)<br>(d) powder | QW-404.23 | | E | E | E | |
| The addition or deletion of filler material | 4.14.3(8) | | | | E | | The deletion or addition of filler metal | QW-404.14 | | | | E | |
| A change from single to multiple supplementary filler metal or vice versa | 4.13.3(12) | E | E | E | E | E | The addition, deletion, or change of more than 10% in the volume of supplemental filler metal | QW-404.24 | | E | E | E | E |
| Supplemental filler metal | 4.13.3(6) | NE | NE | NE | NE | NE | | | | | | | |
| Addition or deletion of supplementary filler metal (powder or wire), or a change of 10% in the amount | 4.14.3(11) | E | E | | | E | | | | | | | |
| Addition or deletion, or a change in the nominal amount or composition of supplementary metal (in addition to filler metal) beyond that qualified | 4.14.3(14) | | E | | | E | | | | | | | |
| Energized filler metal "hot" | 4.13.3(10) | | | | NE | | N/A | | | | | | |

**Table 6.4** Position variables for a corrosion resistant overlay procedure

| AWS B2.1 | | | | | | | ASME IX | | | | | | |
|---|---|---|---|---|---|---|---|---|---|---|---|---|---|
| Variable | Ref. | SMAW | GMAW | FCAW | GTAW | SAW | Variable | Ref. | SMAW | GMAW | FCAW | GTAW | SAW |
| Welding positions(s) | 4.13.4(1) | NE | NE | NE | NE | NE | Except as specified below, the addition of other welding positions than already qualified<br>(a) Qualification in the horizontal, vertical, or overhead position shall also qualify for the flat position. Qualification in the horizontal fixed position, 5G, shall qualify for the flat, vertical, and overhead positions. Qualification in the horizontal, vertical, and overhead positions shall qualify for all positions. Qualification in the inclined fixed position, 6G, shall qualify for all positions<br>(b) An organization who does production welding in a particular orientation may make the tests for procedure qualification in this particular orientation. Such qualifications are valid only for the positions actually tested, except that an angular deviation of $\pm 15$ deg is permitted in the inclination of the weld axis and the rotation of the weld face as defined in Figure QW-461.1. A test specimen shall be taken from the test coupon in each special orientation<br>(c) For hard-facing and corrosion-resistant weld metal overlay, qualification in the 3G, 5G, or 6G positions, where 5G or 6G pipe coupons include at least one vertical segment completed utilizing the up-hill progression or a 3G plate coupon is completed utilizing the up-hill progression, shall qualify for all positions. Chemical analysis, hardness, macro-etch, and at least two of the bend tests, as required in Table QW-453, shall be removed from the vertical uphill overlaid segment as shown in Figure QW-462.5(b)<br>(d) A change from the vertical down to vertical up-hill progression shall require requalification | QW-405.4 | E | E | E | E | E |
| Progression for vertical welding | 4.13.4(2) | NE | NE | NE | NE | NE | | | | | | | |
| The addition of a welding position, except that positions other than flat also qualify for flat | 4.14.4(2) | E | E | E | E | E | | | | | | | |

**Table 6.5** Preheat and interpass temperature variables for a corrosion resistant overlay procedure

| AWS B2.1 | | | | | | | ASME IX | | | | | | |
|---|---|---|---|---|---|---|---|---|---|---|---|---|---|
| Variable | Ref. | SMAW | GMAW | FCAW | GTAW | SAW | Variable | Ref. | SMAW | GMAW | FCAW | GTAW | SAW |
| Preheat minimum | 4.13.5(1) | NE | NE | NE | NE | NE | A decrease of more than 100 °F (55 °C) in the preheat temperature qualified or an increase in the maximum interpass temperature recorded on the PQR. The minimum temperature for welding shall be specified in the WPS | QW-406.4 | E | E | E | E | E |
| A decrease in preheat of more than 100 °F [38 °C] from that qualified | 4.14.5(1) | E | E | E | E | E | | | | | | | |
| For M-23, M-24, M-26, and M-27 heat-treatable materials an increase in the preheat or interpass temperature of more than 100 °F [38 °C] from that qualified | | E | E | | E | | | | | | | | |
| Preheat maintenance | 4.13.5(3) | NE | NE | NE | NE | NE | N/A | | | | | | |

**Table 6.6** Postweld heat treatment variables for a corrosion resistant overlay procedure

| AWS B2.1 | | | | | | | ASME IX | | | | | | |
|---|---|---|---|---|---|---|---|---|---|---|---|---|---|
| Variable | Ref. | SMAW | GMAW | FCAW | GTAW | SAW | Variable | Ref. | SMAW | GMAW | FCAW | GTAW | SAW |
| PWHT temperature and time | 4.13.6(1) | NE | NE | NE | NE | NE | A separate procedure qualification is required for each of the following:<br>(a) For weld corrosion-resistant overlay of A-No. 8 on all base materials, a change in postweld heat treatment condition in QW-407.1, or when the total time at postweld heat treatment encountered in fabrication exceeds 20 h, an increase of 25% or more in total time at postweld heat treating temperature<br>(b) For weld corrosion-resistant overlay of A-No. 9 on all base materials, a change in postweld heat treatment condition in QW-407.1, or an increase of 25% or more in total time at postweld heat treating temperature<br>(c) For all other weld corrosion-resistant overlays on all base materials, a change in postweld heat treatment condition in QW-407.1 | QW-407.9 | E | E | E | E | E |
| For the following M-numbers 1, 3, 4, 5, 6, 7, 9, 10, 11, and 12, a change from any one condition to any other requires requalification:<br>(a) No PWHT<br>(b) PWHT below the lower transformation temperature<br>(c) PWHT within the transformation temperature range<br>(d) PWHT above the upper transformation temperature<br>(e) PWHT above the upper transformation temperature, followed by treatment below the lower transformation temperature | 4.14.6(1) | E | E | E | E | E | | | | | | | |
| For all materials not covered above, a separate PQR is required for no PWHT and PWHT | 4.14.6(2) | E | E | E | E | E | | | | | | | |

**Table 6.7** Shielding gas variables for a corrosion resistant overlay procedure

| AWS B2.1 | | | | | | | ASME IX | | | | | | |
|---|---|---|---|---|---|---|---|---|---|---|---|---|---|
| Variable | Ref. | SMAW | GMAW | FCAW | GTAW | SAW | Variable | Ref. | SMAW | GMAW | FCAW | GTAW | SAW |
| Addition or deletion of torch shielding gas | 4.14.7(1) | | E | E | E | | A separate procedure qualification is required for each of the following:<br>(a) a change from a single shielding gas to any other single shielding gas<br>(b) a change from a single shielding gas to a mixture of shielding gasses, and vice versa<br>(c) a change in the specified percentage composition of a shielding gas mixture<br>(d) the addition or omission of shielding gas<br>The AWS classification of SFA-5.32 may be used to specify the shielding gas composition | QW-408.2 | | E | E | E | |
| A change in the specified nominal composition of shielding gas | 4.14.7(2) | | E | E | E | | | | | | | | |
| Torch shielding gas and flow rate range | 4.13.7(1) | | NE | NE | NE | | A change in the specified flow rate range of the shielding gas or mixture of gases | QW-408.3 | | NE | NE | NE | |
| For M-21 through M-27, an increase of 50% or more, or a decrease of 20% or more in the shielding gas flow rate used for qualification | 4.14.7(7) | | E | | E | | | | | | | | |

**Table 6.8** Electrical characteristics variables for a corrosion resistant overlay procedure

| AWS B2.1 | | | | | | | ASME IX | | | | | | |
|---|---|---|---|---|---|---|---|---|---|---|---|---|---|
| Variable | Ref. | SMAW | GMAW | FCAW | GTAW | SAW | Variable | Ref. | SMAW | GMAW | FCAW | GTAW | SAW |
| Current (or wire feed speed), current type, and polarity | 4.13.8(1) | NE | NE | NE | NE | NE | A change from AC to DC, or vice versa; and in DC welding, a change from electrode negative (straight polarity) to electrode positive (reverse polarity), or vice versa. This variable does not apply to a WPS qualified for welding base metals that are assigned to P-No. 8, P-Nos. 21 through 26, and P-Nos. 41 through 49 | QW-409.4 | E | E | E | E | E |
| | | | | | | | A change in the range of amperage, or except for SMAW, GTAW, or waveform controlled welding, a change in the range of voltage. A change in the range of electrode wire feed speed may be used as an alternative to amperage. See Nonmandatory Appendix H | QW-409.8 | NE | NE | NE | NE | NE |
| N/A | | | | | | | An increase of more than 10% in the amperage used in application for the first layer | QW-409.22 | E | | | | |
| | | | | | | | For the first layer only, an increase in heat input of more than 10% or an increase in volume of weld metal deposited per unit length of weld of more than 10%. The increase shall be determined by the methods of QW-409.1 | QW-409.26 | | E | E | E | E |
| Voltage range (except for manual welding) | 4.13.8(2) | | NE | NE | NE | NE | N/A | | | | | | |
| Transfer mode | 4.13.8(5) | | NE | NE | | | N/A | | | | | | |
| A change to or from pulsed current | 4.13.8(6) | | NE | NE | NE | | N/A | | | | | | |
| Specification, classification, and diameter of tungsten electrode | 4.13.8(4) | | | | NE | | A change in type or size of tungsten electrode | QW-409.12 | | | | NE | |

**Table 6.9** Other variables for a corrosion resistant overlay procedure

| AWS B2.1 | | | | | | | ASME IX | | | | | | |
|---|---|---|---|---|---|---|---|---|---|---|---|---|---|
| Variable | Ref. | SMAW | GMAW | FCAW | GTAW | SAW | Variable | Ref. | SMAW | GMAW | FCAW | GTAW | SAW |
| A change in welding process | 4.14.9(1) | E | E | E | E | E | A change from one welding process to another welding process is an essential variable and requires requalification | QW-401 | E | E | E | E | E |
| Stringer or weave bead | 4.13.9(11) | NE | NE | NE | NE | NE | For manual or semiautomatic welding, a change from the stringer bead technique to the weave bead technique, or vice versa | QW-410.1 | NE | NE | NE | NE | NE |
| A change from a stringer to a weave bead, but not vice versa, for M-23, M-24, M-26, and M-27 heat-treatable materials | 4.14.9(12) | | E | | E | | | | | | | | |
| N/A | | | | | | | A change in the orifice, cup, or nozzle size | QW-410.3 | | NE | NE | NE | |
| Cleaning | 4.13.9(5) | NE | NE | NE | NE | NE | A change in the method of initial and interpass cleaning (brushing, grinding, etc.) | QW-410.5 | NE | NE | NE | NE | NE |
| N/A | | | | | | | For the machine or automatic welding process, a change in width, frequency, or dwell time of oscillation technique | QW-410.7 | | NE | NE | NE | NE |

(continued)

**Table 6.9** (continued)

| AWS B2.1 | | | | | | | ASME IX | | | | | | |
|---|---|---|---|---|---|---|---|---|---|---|---|---|---|
| Variable | Ref. | SMAW | GMAW | FCAW | GTAW | SAW | Variable | Ref. | SMAW | GMAW | FCAW | GTAW | SAW |
| Contact tube to work distance | 4.13.9(4) | | NE | NE | | NE | A change in the contact tube to work distance | QW-410.8 | | NE | NE | | NE |
| For mechanized or automatic, single or multiple electrode and spacing | 4.13.9(2) | | NE | NE | NE | NE | A change in the spacing of multiple electrodes for machine or automatic welding | QW-410.15 | | | | NE | NE |
| Welding process and whether manual, semiautomatic, mechanized, or automatic | 4.13.9(1) | NE | NE | NE | NE | NE | A change from manual or semiautomatic to machine or automatic welding and vice versa | QW-410.25 | | NE | NE | NE | NE |
| Peening | 4.13.9(6) | NE | NE | NE | NE | NE | The addition or deletion of peening | QW-410.26 | NE | NE | NE | NE | NE |
| N/A | | | | | | | A change from multiple-layer to single layer cladding/ hardsurfacing, or vice versa | QW-410.38 | E | E | E | E | E |
| N/A | | | | | | | For submerged-arc welding and electroslag welding, the deletion of a supplementary device for controlling the magnetic field acting on the weld puddle | QW-410.40 | | | | | E |
| N/A | | | | | | | A change in the number of electrodes acting on the same welding puddle | QW-410.50 | | E | E | E | E |
| N/A | | | | | | | A change in the method of delivering the filler metal to the molten pool, such as from the leading or trailing edge of the torch, the sides of the torch, or through the torch | QW-410.52 | | | | NE | |

## 6.2 Examination and Testing Requirements

The testing methods and the number of tests required for a corrosion resistant overlay procedure qualification test for both AWS B2.1 (ref. 4.9 and Table 4.1) and ASME IX (ref. QW-214 and Table QW-453) are summarized in Table 6.10.

For AWS B2.1, mechanical testing specimen blanks shall be removed from the locations shown in AWS B2.1 Fig. 4.8. Note the test location for a chemical analysis specimen should be specified in the referencing document as it is not specified in AWS B2.1.

For ASME IX, mechanical testing specimen blanks shall be removed from locations shown in ASME IX Figure QW-462.5(d) for a plate qualification test weldment and Figure QW-462.5(c) for a pipe qualification test weldment. If chemical analysis is specified by the referencing document, the test locations are shown in ASME IX Figure QW-462.5(e) for a plate qualification test weldment and Figure QW-462.5(b) for a pipe qualification test weldment.

Special test weldments shall be examined and tested as specified by the referencing document. When a test in Table 6.10 is specified by the referencing document, the acceptance criteria shall be as required in AWS B2.1 or ASME IX. The criteria for all other tests shall be as specified in the referencing document.

Depending on the application of the welding procedure, material type(s), and additional requirements or standards invoked by the referencing document, additional testing may be specified. Additional testing may include:

- Ferrite Number Testing
- Corrosion Testing
- Hardness Testing.

*Visual Examination for AWS B2.1*

Visual examination shall be performed if required by the referencing document (ref. AWS B2.1 Table 4.1 note b). Examination procedure and acceptance criteria shall conform to the requirements of the referencing code.

**Table 6.10** Test methods required for corrosion resistant overlay procedure qualification

| | AWS B2.1 | ASME IX |
|---|---|---|
| Visual examination | If specified | Not required |
| Penetrant examination | Yes | Yes |
| Bend testing | Yes (qty. 4) | Yes (qty. 4) |
| Chemical analysis | Yes | Not required |

*Visual Examination for ASME IX*
There is no requirement for visual examination.

*Penetrant Examination for AWS B2.1*
Liquid penetrant examination shall be performed in accordance with ASTM E165, *Standard Test Method for Liquid Penetrant Examination*. The surface shall be evaluated based on the following acceptance criteria (ref. AWS B2.1 4.9.1):

- No linear indications longer than 1/16 in. [1.5 mm].
- No more than four rounded indications in a line with dimensions greater than 1/16 in. [1.5 mm] and separated from each other by less than 1/16 in. [1.5 mm].

*Penetrant Examination for ASME IX*
Liquid penetrant examination shall be performed in accordance with the requirements of ASME BPVC Section V *Nondestructive Examination*, Article 6 (ref. ASME IX QW-195.1 and QW-214.2(a)). The surface shall be evaluated based on the following acceptance criteria (ref. ASME IX QW-195.2):

- No linear indications with major dimensions greater than 1/16 in. [1.5 mm] having a length greater than three times the width.
- No rounded indications of circular or elliptical shape greater than 3/16 in. [5 mm] with the length equal to or less than three times the width.
- No more than four rounded indications of circular or elliptical shape with the length equal to or less than three times the width in a line separated by 1/16 in. [1.5 mm] or less (edge-to-edge).

*Bend Testing for AWS B2.1*
AWS B2.1 allows qualification testing by either transverse or longitudinal side bends. The preparation and dimensions of specimen blanks for side bend testing are provided in AWS B2.1 Annex A Figure A.4A. Bend specimens shall be bent in one of the guided bend test fixtures shown in AWS B2.1 Annex A Figure A.5A bottom ejecting guided bend fixture, Figure A.5B bottom type guided bend fixture, or Figure A.5C wrap around guided bend fixture.

The acceptance criteria for bend testing in AWS B2.1 4.9.4 states:

- There shall be no open discontinuity exceeding 1/16 in. [1.5 mm] in the cladding measured in any direction on the convex surface of the specimen after bending.
- There shall be no open defects exceeding 1/8 in. [3 mm] in length at the weld interface after bending.

*Bend Testing for ASME IX*

ASME IX allows qualification testing by 2 transverse side bends and 2 longitudinal side bends or 4 transverse side bends (ref. ASME IX Table QW-453). The preparation and dimensions of specimen blanks for bend testing are provided in ASME IX QW-161.1 and Figure QW-462.2 for transverse side bend specimens. The test specimens shall be cut so that there are either two specimens parallel and two specimens perpendicular to the direction of the welding, or four specimens perpendicular to the direction of the welding. For coupons that are less than 3/8 in. (10 mm) thick, the width of the side-bend specimens may be reduced to the thickness of the test coupon (ref. ASME IX QW-214.2(b)). Bend specimens shall be bent in one of the guided bend test fixtures shown in ASME IX Figure QW-466.1 test jig dimensions, Figure QW-466.2 guided-bend roller jig, or Figure QW-466.3 guided-bend wrap around jig.

The acceptance criteria for side bend testing in ASME IX QW-163 states:

- There shall be no open discontinuity exceeding 1/16 in. [1.5 mm] in the cladding measured in any direction.
- There shall be no open discontinuity exceeding 1/8 in. [3 mm] along the approximate weld interface.

*Chemical Analysis for AWS B2.1*

AWS B2.1 4.9.5 states a chemical analysis sample shall be removed as shown in AWS B2.1 Figure A.4B. The results from the chemical analysis specimen shall meet the requirements of the referencing document.

*Chemical Analysis for ASME IX*

ASME IX QW-214.2(c) states when a chemical composition is specified by the referencing document, chemical analysis specimens shall be removed at locations specified in ASME IX Figures QW-462.5(b) or QW-462.5(e). The chemical analysis shall be performed in accordance with ASME IX Figure QW-462.5(a) and shall be within the range specified by the referencing document. This chemical analysis is not required when a chemical composition is not specified by the referencing document.

## 6.3 Qualification Limits

In addition to the restrictions of specific variables for the different welding processes as outlined in Table 6.1 through Table 6.8, additional limitations for Position, Base Metal Thickness, and Weld Deposit Thickness are discussed below.

### 6.3.1 Process Control

For corrosion resistant overlay procedure qualifications, both AWS B2.1 and ASME IX allows qualification by any form of process control, whether manual, semiautomatic, mechanized/machine, or automatic, to qualify the welding procedure for all forms of process control (ref. AWS B2.1 4.13.9(1) and ASME IX QW-410.25). The qualified welding procedure may however restrict a change from one form of process control to another.

### 6.3.2 Position

For corrosion resistant overlay weld procedure qualification, both AWS B2.1 and ASME IX limit the qualified positions based on the positions tested as shown in Table 6.11 (ref. AWS B2.1 4.14.4(2) and ASME IX QW-405.4). The standard welding positions (1 Flat, 2 Vertical, 3 Horizontal, 4 Overhead) for corrosion resistant overlay on plate or pipe are defined in AWS B2.1 Figure A.1A and ASME IX Figures QW-461.3 and QW-461.4.

Both AWS B2.1 and ASME IX define a change in position as an angular deviation of>±15 degrees from the specified horizontal and vertical planes, and an angular deviation of>±5 degrees from the specified inclined plane are permitted during welding. This is depicted for corrosion resistant overlay welds in AWS B2.1 Figure A.1B and ASME IX QW-461.1.

### 6.3.3 Base Metal Thickness

The base metal thickness range qualified for corrosion resistant overlay weld procedures (WPS) are governed by the qualification test weldment base metal thickness. The qualified corrosion resistant overlay base metal thickness ranges are identical for AWS B2.1 (ref. Table 4.4) and ASME IX (ref. Table QW-453) and are shown in Table 6.12.

**Table 6.11** Qualified positions for corrosion resistant overlay

| Position(s) tested | | Position(s) qualified | |
|---|---|---|---|
| | | AWS B2.1 | ASME IX |
| Plate | 1 | 1 | 1 |
| | 2 | 1, 2 | 1, 2 |
| | 3 (uphill progression) | 1, 3 (uphill or downhill progression) | 1, 3 (uphill progression) |
| | 3 (downhill progression) | 1, 3 (uphill or downhill progression) | 1, 3 (downhill progression) |
| | 4 | 1, 4 | 1, 4 |
| | 2 + 3 (uphill progression) + 4 | 1, 2, 3 (uphill or downhill progression), 4 | 1, 2, 3 (uphill progression), 4, 5, 6 |
| | 2 + 3 (downhill progression) + 4 | 1, 2, 3 (uphill or downhill progression), 4 | 1, 2, 3 (downhill progression), 4, 5, 6 |
| Pipe | 1 (pipe rotated) | 1 | 1 |
| | 2 (circumferential) | 1, 2 | 1, 2 |
| | 3 (pipe rotated) | 1, 3 | 1, 3 |
| | 4 (pipe rotated) | 1, 4 | 1, 4 |
| | 5 (circumferential) | 1, 5 | 1, 3, 4, 5 |
| | 6 (circumferential) | 1, 6 | 1, 2, 3, 4, 5, 6 |

Notes
(1) AWS appends position with the letter "C" to denote corrosion resistant overlay while ASME does not differentiate position for corrosion resistant overlay
(2) Position of welding:
1 = Flat
2 = Horizontal
3 = Vertical
4 = Overhead
5 = Horizontal Pipe
6 = 45° Inclined Pipe

**Table 6.12** Qualified base metal thickness for corrosion resistant overlay

| Test weldment thickness ($T$) in. [mm] | Base metal thickness qualified | |
|---|---|---|
| | Minimum in. [mm] | Maximum in. [mm] |
| <1 [25] | $T$ | Unlimited |
| ≥1 [25] | 1 [25] | Unlimited |

### 6.3.4 Weld Deposit Thickness

The qualified corrosion resistant overlay weld deposit thickness ranges are different for AWS B2.1 and ASME IX. For AWS B2.1, the minimum weld metal thickness qualified for cladding and hardfacing is one layer if the test weldment has only one layer, and is two layers if the test weldment has two or more layers. The number of layers applies individually to each filler metal classification (ref. AWS B2.1 Table 4.4).

For ASME IX, the minimum weld metal thickness qualified for cladding and hardfacing is the distance from the approximate weld interface to the final as-welded surface as shown in Figures QW-462.5(a) through QW-462.5(e). There is no limit on the maximum thickness for corrosion resistant or hard-facing weld metal overlay that may be used in production (ref. ASME IX QW-402.16).

## 6.4 Summary

This chapter discussed the specific requirements for the qualification of corrosion resistant overlay welding procedures in accordance with the rules of AWS B2.1 and ASME IX. The required welding variables to be documented in the Procedure Qualification Record (PQR) are discussed for each welding process. The required examination and test methods along with their acceptance criteria are also outlined. Finally, the required welding variables to be addressed in the Welding Procedure Specification (WPS) along with their qualified limits and ranges are discussed.

# 7 Procedure Qualification for Hardfacing Overlay

The concepts of developing a qualified welding procedure and producing the required welding documentation were introduced in Chap. 4. This chapter discusses the specific requirements of documenting the welding variables utilized the Procedure Qualification Record (PQR) and the Welding Procedure Specification (WPS).

## 7.1 Hardfacing Overlay Variables

The following tables illustrate the required welding variables for hardfacing overlay procedures qualified under AWS B2.1 and ASME BPVC IX. The essential (E), supplementary essential (SE), and non-essential (NE) variables are compared for the five most common arc welding procedures (SMAW, GMAW, FCAW, GTAW, SAW). Variables mark N/A are not addressed for a given standard. The tables should be regarded as illustrative only since the various standards are revised periodically.

© The Author(s), under exclusive license to Springer Nature Switzerland AG 2025

D. Barborak, *Arc Welding Qualification Standards*, Synthesis Lectures on Welding Engineering, https://doi.org/10.1007/978-3-031-64646-1_7

### 7.1.1 Joint Design

See Table 7.1.

### 7.1.2 Base Metal

See Table 7.2.

### 7.1.3 Filler Metals

See Table 7.3.

### 7.1.4 Position

See Table 7.4.

### 7.1.5 Preheat and Interpass Temperature

See Table 7.5.

### 7.1.6 Postweld Heat Treatment

See Table 7.6.

### 7.1.7 Shielding Gas

See Table 7.7.

### 7.1.8 Electrical Characteristics

See Table 7.8.

### 7.1.9 Other Variables

See Table 7.9.

**Table 7.1** Joint design variables for a hardfacing overlay procedure

| AWS B2.1 | | | | | | | ASME BPVC.IX | | | | | | |
|---|---|---|---|---|---|---|---|---|---|---|---|---|---|
| Variable | Ref. | SMAW | GMAW | FCAW | GTAW | SAW | Variable | Ref. | SMAW | GMAW | FCAW | GTAW | SAW |
| A decrease in thickness or a change in the nominal chemical composition of surfacing or buttering beyond that qualified. (Note: this variable is listed under the filler metal grouping) | 4.14.3(9) | E | E | E | E | E | A decrease in the distance between the approximate weld interface and the final surface of the production corrosion-resistant or hard-facing weld metal overlay below the minimum thickness qualified as shown in Figures QW-462.5(a) through QW-462.5(e). There is no limit on the maximum thickness for corrosion resistant or hard-facing weld metal overlay that may be used in production | QW-402.16 | E | E | E | E | E |

**Table 7.2** Base metal variables for a hardfacing overlay procedure

| AWS B2.1 | | | | | | | ASME BPVC.IX | | | | | | |
|---|---|---|---|---|---|---|---|---|---|---|---|---|---|
| Variable | Ref. | SMAW | GMAW | FCAW | GTAW | SAW | Variable | Ref. | SMAW | GMAW | FCAW | GTAW | SAW |
| M-number and group number | 4.13.2(1) | NE | NE | NE | NE | NE | If the chemical composition of the weld metal overlay is specified in the WPS, a change in the P-number listed in Table QW/QB-422 to another P-number or unlisted base metal, or a change in Group Number for P-No. 10 or P-No. 11 base metals<br>If the chemical composition of the weld metal overlay is not specified in the WPS, qualification on P-No. 5A or any lower P-number base metal also qualifies for weld metal overlay on all lower P-number base metals | QW-403.20 | E | E | E | E | E |
| Cladding and hardfacing require separate qualification for each base metal M-number, and filler metal combination | 4.3.17 | E | E | E | E | E | | | | | | | |
| A change from one M-number base metal to another M-number base metal or to a combination of M-number base metals except as permitted in 4.3.10 | 4.14.2(2) | E | E | E | E | E | | | | | | | |
| A change from one unlisted base metal to a different unlisted base metal or to a listed base metal or vice versa | 4.14.2(3) | E | E | E | E | E | | | | | | | |
| A change from one M-5 group (A, B, etc.) to any other. A change from M-9A to M-9B, but not vice versa. A change from M-10 to M-11 group (A, B, etc.) to any other group | 4.14.2(5) | E | E | E | E | E | | | | | | | |

(continued)

**Table 7.2** (continued)

| AWS B2.1 | | | | | | | ASME BPVC.IX | | | | | | |
|---|---|---|---|---|---|---|---|---|---|---|---|---|---|
| Variable | Ref. | SMAW | GMAW | FCAW | GTAW | SAW | Variable | Ref. | SMAW | GMAW | FCAW | GTAW | SAW |
| Thickness range qualified | 4.13.2(2) | NE | NE | NE | NE | NE | A change in base metal thickness beyond the range qualified in Table QW-453 | QW-403.23 | E | E | E | E | E |
| A change in base metal thickness beyond the range permitted in 4.5 | 4.14.2(1) | E | E | E | E | E | | | | | | | |
| Limitations on the thickness ranges qualified by procedure qualification tests are given in the following tables:<br>Table 4.4—thickness limitations for cladding and hardfacing for procedure qualification | 4.5.1 | E | E | E | E | E | | | | | | | |
| The coating description or type, if present | 4.13.2(4) | NE | NE | NE | NE | NE | N/A | | | | | | |
| A change from an uncoated metal to a metal coated (such as painted or galvanized) metal unless the coating is removed from the weld area prior to welding, but not vice versa, as permitted in 4.3.8 | 4.14.2(6) | E | E | E | E | E | | | | | | | |

**Table 7.3** Filler metals variables for a hardfacing overlay procedure

| AWS B2.1 | | | | | | | ASME BPVC.IX | | | | | | |
|---|---|---|---|---|---|---|---|---|---|---|---|---|---|
| Variable | Ref. | SMAW | GMAW | FCAW | GTAW | SAW | Variable | Ref. | SMAW | GMAW | FCAW | GTAW | SAW |
| Specification, classification, F- and A-number, or if not classified the nominal composition | 4.13.1(1) | NE | NE | NE | NE | NE | A change in the filler metal classification within an SFA specification, or for a filler metal not covered by an SFA *specification or a* filler metal with a "G" suffix within an SFA specification, a change in the trade name of the filler metal | QW-404.12 | E | E | E | E | E |
| Specification, classification, F- and A-number, or if not classified the nominal composition | 4.13.1(1) | NE | NE | NE | NE | NE | A change in the composition of the deposited weld metal from one A-number in Table QW-442 to any other A-number, or to an analysis not listed in the table. A change in the UNS number for each AWS classification of A-No. 8 or A-No. 9 analysis of Table QW-442, or each nonferrous alloy in Table QW-432, shall require separate WPS qualification. A-numbers may be determined in accordance with QW-404.5 | QW-404.37 | | E | E | | |
| A change from one F-number to any other F-number or to any filler metal not listed in Annex B | 4.14.3(1) | E | E | E | E | E | | | | | | | |
| For ferrous materials, a change from one A-number to any other A-number or to a filler metal analysis not listed in Annex B (the PQR and WPS shall state the nominal chemical composition, the AWS specification and classification, or the manufacturer's designation for filler metals which do not fall in an A-number group). Qualification with A-1 shall qualify for A-2 and vice versa | 4.14.3(2) | E | E | E | E | E | | | | | | | |

(continued)

**Table 7.3** (continued)

| AWS B2.1 | | | | | | | ASME BPVC.IX | | | | | | |
|---|---|---|---|---|---|---|---|---|---|---|---|---|---|
| Variable | Ref. | SMAW | GMAW | FCAW | GTAW | SAW | Variable | Ref. | SMAW | GMAW | FCAW | GTAW | SAW |
| For surfacing, a change in the chemical composition of the weld metal (A-number or alloy type). Each layer shall be considered independent of other layers | 4.14.3(3) | E | E | E | E | E | | | | | | | |
| A decrease in thickness or a change in the nominal chemical composition of surfacing or buttering beyond that qualified | 4.14.3(9) | E | E | E | E | E | | | | | | | |
| If the weld metal alloy content is largely dependent upon the composition of the flux, any change in the welding procedure which would result in the important weld metal alloying elements being outside the specified chemical composition range of the WPS | 4.14.3(6) | | | | | E | Where the alloy content of the weld metal is largely dependent upon the composition of the supplemental filler metal (including powder filler metal for PAW), any change in any part of the welding procedure that would result in the important alloying elements in the weld metal being outside of the specification range of chemistry given in the welding procedure specification | QW-404.27 | | E | E | E | E |

(continued)

**Table 7.3** (continued)

| AWS B2.1 | | | | | | | ASME BPVC.IX | | | | | | |
|---|---|---|---|---|---|---|---|---|---|---|---|---|---|
| Variable | Ref. | SMAW | GMAW | FCAW | GTAW | SAW | Variable | Ref. | SMAW | GMAW | FCAW | GTAW | SAW |
| Flux-electrode classification | 4.13.3(4) | | | | | NE | For submerged-arc welding and electroslag welding, a change in the nominal composition or type of flux used. Requalification is not required for a change in flux particle size | QW-404.39 | | | | | E |
| A change from one AWS flux-electrode classification listed to any other electrode flux-electrode classification, or to an unlisted electrode-flux classification. A variation of 0.5% of the molybdenum content of the weld metal does not require requalification | 4.14.3(16) | | | | | E | | | | | | | |
| Weld metal thickness by process and filler metal classification | 4.13.3(2) | NE | NE | NE | NE | NE | N/A | | | | | | |
| A change in the weld metal thickness beyond that permitted in 4.5 | 4.14.3(17) | E | E | E | E | E | | | | | | | |
| Filler metal size or diameter<br>A change of filler metal/electrode nominal size/shape in the first layer | 4.14.3(3)<br>4.14.3(10) | NE<br>E | NE<br>E | NE<br>E | | NEE | A change in the nominal electrode diameter used for the first layer of deposit | QW-404.38 | NE | | | | |
| | | | | | | | A change in the nominal size of the electrode or electrodes specified in the WPS | QW-404.6 | | NE | NE | | NE |
| | | | | | | | A change in the size of the filler metal | QW-404.3 | | | | NE | |

(continued)

**Table 7.3** (continued)

| AWS B2.1 | | | | | | | ASME BPVC.IX | | | | | | |
|---|---|---|---|---|---|---|---|---|---|---|---|---|---|
| Variable | Ref. | SMAW | GMAW | FCAW | GTAW | SAW | Variable | Ref. | SMAW | GMAW | FCAW | GTAW | SAW |
| A change from wire to strip electrodes and vice versa | 4.14.3(15) | | | | | E | An increase in the nominal thickness or width of the electrode for strip filler metals used with the SAW and ESW processes for corrosion-resistant and hard-facing weld metal overlay | QW-404.57 | | | | | E |
| N/A | | | | | | | A change from one of the following filler metal product forms to another:<br>(a) bare (solid or metal cored)<br>(b) flux cored<br>(c) flux coated (solid or metal cored)<br>(d) powder | QW-404.23 | | E | E | E | |
| The addition or deletion of filler material | 4.14.3(8) | | | | E | | The deletion or addition of filler metal | QW-404.14 | | | | E | |
| A change from single to multiple supplementary filler metal or vice versa | 4.13.3(12) | E | E | E | E | E | The addition, deletion, or change of more than 10% in the volume of supplemental filler metal | QW-404.24 | | E | E | | E |
| Supplemental filler metal | 4.13.3(6) | NE | NE | NE | NE | NE | | | | | | | |
| Addition or deletion of supplementary filler metal (powder or wire), or a change of 10% in the amount | 4.14.3(11) | E | E | | | E | | | | | | | |

(continued)

**Table 7.3** (continued)

| AWS B2.1 | | | | | | | ASME BPVC.IX | | | | | | |
|---|---|---|---|---|---|---|---|---|---|---|---|---|---|
| Variable | Ref. | SMAW | GMAW | FCAW | GTAW | SAW | Variable | Ref. | SMAW | GMAW | FCAW | GTAW | SAW |
| Addition or deletion, or a change in the nominal amount or composition of supplementary metal (in addition to filler metal) beyond that qualified | 4.14.3(14) | | E | | | E | | | | | | | |
| Energized filler metal "hot" | 4.13.3(10) | | | | NE | | N/A | | | | | | |

**Table 7.4** Position variables for a hardfacing overlay procedure

| AWS B2.1 | | | | | | | ASME BPVC.IX | | | | | | |
|---|---|---|---|---|---|---|---|---|---|---|---|---|---|
| Variable | Ref. | SMAW | GMAW | FCAW | GTAW | SAW | Variable | Ref. | SMAW | GMAW | FCAW | GTAW | SAW |
| Welding positions(s) | 4.13.4(1) | NE | NE | NE | NE | NE | Except as specified below, the addition of other welding positions than already qualified. (a) Qualification in the horizontal, vertical, or overhead position shall also qualify for the flat position. Qualification in the horizontal fixed position, 5G, shall qualify for the flat, vertical, and overhead positions. Qualification in the horizontal, vertical, and overhead positions shall qualify for all positions. Qualification in the inclined fixed position, 6G, shall qualify for all positions<br>(b) An organization who does production welding in a particular orientation may make the tests for procedure qualification in this particular orientation. Such qualifications are valid only for the positions actually tested, except that an angular deviation of ±15 deg is permitted in the inclination of the weld axis and the rotation of the weld face as defined in Figure QW-461.1. A test specimen shall be taken from the test coupon in each special orientation<br>(c) For hard-facing and corrosion-resistant weld metal overlay, qualification in the 3G, 5G, or 6G positions, where 5G or 6G pipe coupons include at least one vertical segment completed utilizing the up-hill progression or a 3G plate coupon is completed utilizing the up-hill progression, shall qualify for all positions. Chemical analysis, hardness, macro-etch, and at least two of the bend tests, as required in Table QW-453, shall be removed from the vertical uphill overlaid segment as shown in Figure QW-462.5(b)<br>(d) A change from the vertical down to vertical up-hill progression shall require requalification | QW-405.4 | E | E | E | E | E |
| Progression for vertical welding | 4.13.4(2) | NE | NE | NE | NE | NE | | | | | | | |
| The addition of a welding position, except that positions other than flat also qualify for flat | 4.14.4(2) | E | E | E | E | E | | | | | | | |

**Table 7.5** Preheat and interpass temperature variables for a hardfacing overlay procedure

| AWS B2.1 | | | | | | | ASME BPVC.IX | | | | | | |
|---|---|---|---|---|---|---|---|---|---|---|---|---|---|
| Variable | Ref. | SMAW | GMAW | FCAW | GTAW | SAW | Variable | Ref. | SMAW | GMAW | FCAW | GTAW | SAW |
| Preheat minimum | 4.13.5(1) | NE | NE | NE | NE | NE | A decrease of more than 100 °F (55 °C) in the preheat temperature qualified or an increase in the maximum interpass temperature recorded on the PQR. The minimum temperature for welding shall be specified in the WPS | QW-406.4 | E | E | E | E | E |
| A decrease in preheat of more than 100 °F [38 °C] from that qualified | 4.14.5(1) | E | E | E | E | E | | | | | | | |
| For M-23, M-24, M-26, and M-27 heat-treatable materials an increase in the preheat or interpass temperature of more than 100 °F [38 °C] from that qualified | | E | E | | E | | | | | | | | |
| Preheat maintenance | 4.13.5(3) | NE | NE | NE | NE | NE | N/A | | | | | | |

**Table 7.6** Postweld heat treatment variables for a hardfacing overlay procedure

| AWS B2.1 | | | | | | | ASME BPVC.IX | | | | | | |
|---|---|---|---|---|---|---|---|---|---|---|---|---|---|
| Variable | Ref. | SMAW | GMAW | FCAW | GTAW | SAW | Variable | Ref. | SMAW | GMAW | FCAW | GTAW | SAW |
| PWHT temperature and time | 4.13.6(1) | NE | NE | NE | NE | NE | A change in postweld heat treatment condition in QW-407.1 or an increase of 25% or more in total time at postweld heat treating temperature | QW-407.6 | E | E | E | E | E |
| For the following M-numbers 1, 3, 4, 5, 6, 7, 9, 10, 11, and 12, a change from any one condition to any other requires requalification:<br>(a) No PWHT<br>(b) PWHT below the lower transformation temperature<br>(c) PWHT within the transformation temperature range<br>(d) PWHT above the upper transformation temperature<br>(e) PWHT above the upper transformation temperature, followed by treatment below the lower transformation temperature | 4.14.6(1) | E | E | E | E | E | | | | | | | |
| For all materials not covered above, a separate PQR is required for no PWHT and PWHT | 4.14.6(2) | E | E | E | E | E | | | | | | | |

**Table 7.7** Shielding gas variables for a hardfacing overlay procedure

| AWS B2.1 | | | | | | | ASME BPVC.IX | | | | | | |
|---|---|---|---|---|---|---|---|---|---|---|---|---|---|
| Variable | Ref. | SMAW | GMAW | FCAW | GTAW | SAW | Variable | Ref. | SMAW | GMAW | FCAW | GTAW | SAW |
| Addition or deletion of torch shielding gas | 4.14.7(1) | | E | E | E | | A separate procedure qualification is required for each of the following:<br>(a) a change from a single shielding gas to any other single shielding gas<br>(b) a change from a single shielding gas to a mixture of shielding gasses, and vice versa<br>(c) a change in the specified percentage composition of a shielding gas mixture<br>(d) the addition or omission of shielding gas<br>The AWS classification of SFA-5.32 may be used to specify the shielding gas composition | QW-408.2 | | E | E | E | |
| A change in the specified nominal composition of shielding gas | 4.14.7(2) | | E | E | E | | | | | | | | |

(continued)

**Table 7.7** (continued)

| AWS B2.1 | | | | | | | ASME BPVC.IX | | | | | | |
|---|---|---|---|---|---|---|---|---|---|---|---|---|---|
| Variable | Ref. | SMAW | GMAW | FCAW | GTAW | SAW | Variable | Ref. | SMAW | GMAW | FCAW | GTAW | SAW |
| Torch shielding gas and flow rate range | 4.13.7(1) | | NE | NE | NE | | A change in the specified flow rate range of the shielding gas or mixture of gases | QW-408.3 | | NE | NE | NE | |
| For M-21 through M-27, an increase of 50% or more, or a decrease of 20% or more in the shielding gas flow rate used for qualification | 4.14.7(7) | | E | | E | | | | | | | | |

**Table 7.8** Electrical characteristics variables for a hardfacing overlay procedure

| AWS B2.1 | | | | | | | ASME BPVC.IX | | | | | | |
|---|---|---|---|---|---|---|---|---|---|---|---|---|---|
| Variable | Ref. | SMAW | GMAW | FCAW | GTAW | SAW | Variable | Ref. | SMAW | GMAW | FCAW | GTAW | SAW |
| Current (or wire feed speed), current type, and polarity | 4.13.8(1) | NE | NE | NE | NE | NE | A change from AC to DC, or vice versa; and in DC welding, a change from electrode negative (straight polarity) to electrode positive (reverse polarity), or vice versa. This variable does not apply to a WPS qualified for welding base metals that are assigned to P-No. 8, P-Nos. 21 through 26, and P-Nos. 41 through 49 | QW-409.4 | E | E | E | E | E |
| | | | | | | | A change in the range of amperage, or except for SMAW, GTAW, or waveform controlled welding, a change in the range of voltage. A change in the range of electrode wire feed speed may be used as an alternative to amperage. See nonmandatory Appendix H | QW-409.8 | NE | NE | NE | NE | NE |
| N/A | | | | | | | An increase of more than 10% in the amperage used in application for the first layer | QW-409.22 | E | | | | |

(continued)

**Table 7.8** (continued)

| AWS B2.1 | | | | | | | ASME BPVC.IX | | | | | | |
|---|---|---|---|---|---|---|---|---|---|---|---|---|---|
| Variable | Ref. | SMAW | GMAW | FCAW | GTAW | SAW | Variable | Ref. | SMAW | GMAW | FCAW | GTAW | SAW |
| | | | | | | | For the first layer only, an increase in heat input of more than 10% or an increase in volume of weld metal deposited per unit length of weld of more than 10%. The increase shall be determined by the methods of QW-409.1 | QW-409.26 | | E | E | E | E |
| Voltage range (except for manual welding) | 4.13.8(2) | | NE | NE | NE | NE | N/A | | | | | | |
| Transfer mode | 4.13.8(5) | | NE | NE | | | N/A | | | | | | |
| A change to or from pulsed current | 4.13.8(6) | | NE | NE | NE | | N/A | | | | | | |
| Specification, classification, and diameter of tungsten electrode | 4.13.8(4) | | | | NE | | A change in type or size of tungsten electrode | QW-409.12 | | | | NE | |

**Table 7.9** Other variables for a hardfacing overlay procedure

| AWS B2.1 | | | | | | | ASME BPVC.IX | | | | | | |
|---|---|---|---|---|---|---|---|---|---|---|---|---|---|
| Variable | Ref. | SMAW | GMAW | FCAW | GTAW | SAW | Variable | Ref. | SMAW | GMAW | FCAW | GTAW | SAW |
| A change in welding process | 4.14.9(1) | E | E | E | E | E | A change from one welding process to another welding process is an essential variable and requires requalification | QW-401 | E | E | E | E | E |
| A change exceeding ± 10% in travel speed for mechanized or automatic welding | 4.14.9(10) | | E | E | E | E | N/A | | | | | | |
| A change from stringer bead to weave bead for manual welding | 4.14.9(11) | E | E | E | E | E | For manual or semiautomatic welding, a change from the stringer bead technique to the weave bead technique, or vice versa | QW-410.1 | NE | NE | NE | NE | NE |
| A change from a stringer to a weave bead, but not vice versa, for M-23, M-24, M-26, and M-27 heat-treatable materials | 4.14.9(12) | | E | | E | | | | | | | | |
| N/A | | | | | | | A change in the orifice, cup, or nozzle size | QW-410.3 | | NE | NE | NE | |
| Cleaning | 4.13.9(5) | NE | NE | NE | NE | NE | A change in the method of initial and interpass cleaning (brushing, grinding, etc.) | QW-410.5 | NE | NE | NE | NE | NE |
| N/A | | | | | | | For the machine or automatic welding process, a change in width, frequency, or dwell time of oscillation technique | QW-410.7 | | NE | NE | NE | NE |
| Contact tube to work distance | 4.13.9(4) | | NE | NE | | NE | A change in the contact tube to work distance | QW-410.8 | | NE | NE | | NE |

(continued)

**Table 7.9** (continued)

| AWS B2.1 | | | | | | | ASME BPVC.IX | | | | | | |
|---|---|---|---|---|---|---|---|---|---|---|---|---|---|
| Variable | Ref. | SMAW | GMAW | FCAW | GTAW | SAW | Variable | Ref. | SMAW | GMAW | FCAW | GTAW | SAW |
| For mechanized or automatic, single or multiple electrode and spacing | 4.13.9(2) | | NE | NE | NE | NE | A change in the spacing of multiple electrodes for machine or automatic welding | QW-410.15 | | | | NE | NE |
| Welding process and whether manual, semiautomatic, mechanized, or automatic | 4.13.9(1) | NE | NE | NE | NE | NE | A change from manual or semiautomatic to machine or automatic welding and vice versa | QW-410.25 | | NE | NE | NE | NE |
| Peening | 4.13.9(6) | NE | NE | NE | NE | NE | The addition or deletion of peening | QW-410.26 | NE | NE | NE | NE | NE |
| N/A | | | | | | | A change from multiple-layer to single layer cladding/ hardsurfacing, or vice versa | QW-410.38 | E | E | E | E | E |
| N/A | | | | | | | A change in the number of electrodes acting on the same welding puddle | QW-410.50 | | E | E | E | E |
| N/A | | | | | | | A change in the method of delivering the filler metal to the molten pool, such as from the leading or trailing edge of the torch, the sides of the torch, or through the torch | QW-410.52 | | | | NE | |

## 7.2 Examination and Testing Requirements

The testing methods and the number of tests required for a corrosion resistant overlay procedure qualification test for both AWS B2.1 (ref. 4.10 and Table 4.1) and ASME IX (ref. QW-216 and Table QW-453) are summarized in Table 7.10.

Special test weldments shall be examined and tested as specified by the referencing document. When a test in Table 7.10 is specified by the referencing document, the acceptance criteria shall be as required in AWS B2.1 or ASME IX. The criteria for all other tests shall be as specified in the referencing document.

Depending on the application of the welding procedure, material type(s), and additional requirements or standards invoked by the referencing document, additional testing may be specified. Examples of additional testing may include wear testing and corrosion testing.

*Visual Examination for AWS B2.1*

Visual examination shall be performed if required by the referencing document (ref. AWS B2.1 Table 4.1 note b). Examination procedure and acceptance criteria shall conform to the requirements of the referencing code.

*Visual Examination for ASME IX*

There is no requirement for visual examination.

*Penetrant Examination for AWS B2.1*

Penetrant examination shall be performed if required by the referencing document (ref. AWS B2.1 Table 4.1 note b). Examination procedure and acceptance criteria shall conform to the requirements of the referencing code.

*Penetrant Examination for ASME IX*

Liquid penetrant examination shall be performed and the surface shall be evaluated based on the following acceptance criteria (ref. ASME IX QW-195.2):

- No linear indications with major dimensions greater than 1/16 in. [1.5 mm] having a length greater than three times the width.

**Table 7.10** Test methods required for hardfacing overlay procedure qualification

| | AWS B2.1 | ASME IX |
|---|---|---|
| Visual examination | If specified | Not required |
| Penetrant examination | If specified | Yes |
| Hardness testing | Yes (qty. 3) | Yes (qty. 3) |
| Macroetch examination | If specified | Yes (qty. 2) |
| Chemical analysis | If specified | If specified |

- No rounded indications of circular or elliptical shape greater than 3/16 in. [5 mm] with the length equal to or less than three times the width.
- No more than four rounded indications of circular or elliptical shape with the length equal to or less than three times the width in a line separated by 1/16 in. [1.5 mm] or less (edge-to-edge).

*Hardness Testing for AWS B2.1*

The hardfaced surface shall be prepared for hardness testing at the minimum weld metal thickness to be qualified. Hardness testing shall be performed at three locations, and the result of each test shall not be less than the minimum hardness specified in the referencing document (ref. AWS B2.1 4.10.1).

*Hardness Testing for ASME IX*

After surface conditioning to the minimum thickness specified in the WPS, a minimum of three hardness readings shall be made on each of the specimens from the locations shown in ASMI IX Figure QW-462.5(b) or Figure QW-462.5(e) (ref. QW-216.2(b)). All readings shall meet the requirements of the WPS or referencing document.

*Macro Examination for AWS B2.1*

If macroetch is required by the referencing document, the test weldment shall then be sectioned as shown in Fig. 4.9 (ref. AWS B2.1 4.10.2). Both exposed faces shall then be polished and etched to provide a clear definition of the weld metal and the heat-affected zone in accordance with Annex G of AWS B2.1. Examination results from both faces shall meet the acceptance criteria of the referencing document.

*Macro Examination for ASME IX*

The base metal shall be sectioned transversely to the direction of the hard-facing overlay. The two faces of the hard facing exposed by sectioning shall be polished and etched with a suitable etchant and shall be visually examined with 5X magnification for cracks in the base metal or the heat-affected zone, lack of fusion, or other linear defects (ref. ASME IX QW-216.2(c)). The overlay and base metal shall meet the requirements specified in the WPS or reference document. All exposed faces shall be examined. See ASME IX Figure QW-462.5(b) for pipe and Figure QW-462.5(e) for plate.

*Chemical Analysis for AWS B2.1*

If specified by the referencing document, a chemical analysis sample shall be removed as shown in AWS B2.1 Figure A.4B (ref. AWS B2.1 4.10.3). The results from the chemical analysis specimen shall meet the requirements of the referencing document.

*Chemical Analysis for ASME IX*

When a chemical composition is specified by the referencing document, chemical analysis specimens shall be removed at locations specified in ASME IX Figures QW-462.5(b) or QW-462.5(e) (ref. ASME IX QW-216.2(c)). The chemical analysis shall be performed in accordance with ASME IX Figure QW-462.5(a) and shall be within the range specified by the referencing document. This chemical analysis is not required when a chemical composition is not specified by the referencing document.

If chemical analysis is specified by the referencing document, the test locations are shown in ASME IX Figure QW-462.5(e) for a plate qualification test weldment and Figure QW-462.5(b) for a pipe qualification test weldment.

## 7.3 Qualification Limits

In addition to the restrictions of specific variables for the different welding processes as outlined in Table 7.1 through Table 7.9, additional limitations for position, base metal thickness, and weld deposit thickness are discussed below.

### 7.3.1 Process Control

For corrosion resistant overlay procedure qualifications, both AWS B2.1 and ASME IX allows qualification by any form of process control, whether manual, semiautomatic, mechanized/machine, or automatic, to qualify the welding procedure for all forms of process control (ref. AWS B2.1 4.13.9(1) and ASME IX QW-410.25). The qualified welding procedure may however restrict a change from one form of process control to another.

### 7.3.2 Position

For hardfacing overlay weld procedure qualification, both AWS B2.1 and ASME IX limit the qualified positions based on the positions tested as shown in Table 7.11 (ref. AWS B2.1 4.14.4(2) and ASME IX QW-405.4). The standard welding positions (1 Flat, 2 Vertical, 3 Horizontal, 4 Overhead) for hardfacing overlay on plate or pipe are defined in AWS B2.1 Figure A.1A and ASME IX Figures QW-461.3 and QW-461.4.

Both AWS B2.1 and ASME IX define a change in position as an angular deviation of $>\pm 15$ degrees from the specified horizontal and vertical planes, and an angular deviation of $>\pm 5$ degrees from the specified inclined plane are permitted during welding. This is depicted for corrosion resistant overlay welds in AWS B2.1 Figure A.1B and ASME IX QW-461.1.

**Table 7.11** Qualified positions for hardfacing overlay

| Position(s) tested | | Position(s) qualified | |
|---|---|---|---|
| | | AWS B2.1 | ASME IX |
| Plate | 1 | 1 | 1 |
| | 2 | 1, 2 | 1, 2 |
| | 3 (uphill progression) | 1, 3 (uphill or downhill progression) | 1, 3 (uphill progression) |
| | 3 (downhill progression) | 1, 3 (uphill or downhill progression) | 1, 3 (downhill progression) |
| | 4 | 1, 4 | 1, 4 |
| | 2 + 3 (uphill progression) + 4 | 1, 2, 3 (uphill or downhill progression), 4 | 1, 2, 3 (uphill progression), 4, 5, 6 |
| | 2 + 3 (downhill progression) + 4 | 1, 2, 3 (uphill or downhill progression), 4 | 1, 2, 3 (downhill progression), 4, 5, 6 |
| Pipe | 1 (pipe rotated) | 1 | 1 |
| | 2 (circumferential) | 1, 2 | 1, 2 |
| | 3 (pipe rotated) | 1, 3 | 1, 3 |
| | 4 (pipe rotated) | 1, 4 | 1, 4 |
| | 5 (circumferential) | 1, 5 | 1, 3, 4, 5 |
| | 6 (circumferential) | 1, 6 | 1, 2, 3, 4, 5, 6 |

*Notes*

(1) Neither AWS B2.1 nor ASME IX differentiates positions for hardfacing overlay from groove or fillet welds

(2) Position of welding:

1 = Flat
2 = Horizontal
3 = Vertical
4 = Overhead
5 = Horizontal pipe
6 = 45° Inclined pipe

### 7.3.3 Base Metal Thickness

The base metal thickness range qualified for hardfacing overlay weld procedures (WPS) are governed by the qualification test weldment base metal thickness. The qualified hardfacing overlay base metal thickness ranges are identical for AWS B2.1 (ref. Table 4.4) and ASME IX (ref. Table QW-453) and are shown in Table 7.12.

**Table 7.12** Qualified base metal thickness for hardfacing overlay

| Test weldment thickness ($T$) in. [mm] | Base metal thickness qualified | |
|---|---|---|
| | Minimum in. [mm] | Maximum in. [mm] |
| <1 [25] | $T$ | Unlimited |
| ≥1 [25] | 1 [25] | Unlimited |

### 7.3.4 Weld Deposit Thickness

The qualified hardfacing overlay weld deposit thickness ranges are different for AWS B2.1 and ASME IX. For AWS B2.1, the minimum weld metal thickness qualified for cladding and hardfacing is one layer if the test weldment has only one layer, and is two layers if the test weldment has two or more layers. The number of layers applies individually to each filler metal classification (ref. AWS B2.1 Table 4.4).

For ASME IX, the minimum weld metal thickness qualified for cladding and hardfacing is the distance from the approximate weld interface to the final as-welded surface as shown in Figures QW-462.5(a) through QW-462.5(e). There is no limit on the maximum thickness for corrosion resistant or hard-facing weld metal overlay that may be used in production (ref. ASME IX QW-402.16).

## 7.4 Summary

This chapter discussed the specific requirements for the qualification of hardfacing overlay welding procedures in accordance with the rules of AWS B2.1 and ASME IX. The required welding variables to be documented in the Procedure Qualification Record (PQR) are discussed for each welding process. The required examination and test methods along with their acceptance criteria are also outlined. Finally, the required welding variables to be addressed in the Welding Procedure Specification (WPS) along with their qualified limits and ranges are discussed.

# Performance Qualification

# 8

The concepts of qualifying welding procedures for groove, fillet, corrosion resistant overlay, and hardfacing overlay were discussed in the previous chapters. This chapter discusses the requirements for the performance qualification of a welder or welding operator.

## 8.1 The Performance Qualification Process

The process of conducting a welder or welding operator performance qualification is more simplistic than the process for performing a welding procedure qualification outlined above. The general workflow of the process is shown in Fig. 8.1. Depending on the requirements of the welder/operator in production, the process may be as simple as conducting a single performance qualification test coupon to as complex as conducting multiple performance qualification test coupons. Details for each step of the process are discussed separately below.

### 8.1.1 Requirements Review

Similar to the procedure qualification process, prior to initiating the performance qualification process, all the requirements should be reviewed in preparation for the qualification test. In general, the requirements can be classified as Application specific Requirements, Qualification Requirements, or Other Considerations. These requirements, re-summarized in Table 8.1, are discussed in detail in the Sect. 4.1.

© The Author(s), under exclusive license to Springer Nature Switzerland AG 2025

D. Barborak, *Arc Welding Qualification Standards*, Synthesis Lectures on Welding Engineering, https://doi.org/10.1007/978-3-031-64646-1_8

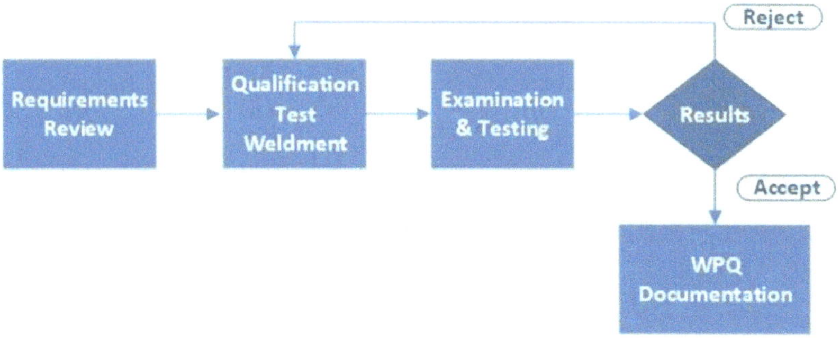


**Fig. 8.1** Workflow of the welder/welding operator performance qualification process

**Table 8.1** Performance qualification requirements

| Application requirements | Qualification requirements | Other considerations |
|---|---|---|
| • Production weld type<br>• Production welding position<br>• Production weldment Thickness<br>• Welding process<br>• Base materials<br>• Welding consumables | • Governing qualification standards<br>• Customer/contractual/jurisdictional requirements<br>• Welding variables<br>• Testing and examination requirements<br>• Test weldment type | • Resource constraints<br>• Weld induced distortion<br>• Welding sequence |

### 8.1.2 Qualification Test Weldment

Both AWS B2.1 and ASME IX allow the performance qualification test coupon for groove, fillet, corrosion resistant overlay, or hardfacing overlay welds to be plate, pipe, or other product forms (ref AWS B2.1 5.1.4 and ASME IX QG-109.2). Both also allow the qualification to be performed on a standard test coupon or a production weldment also known as a "workmanship test". The dimensions of a standard test coupon for performance qualification are typically smaller than those required for procedure qualification due to the reduced number of test specimens required.

### 8.1.3 Examination and Testing

Once welding of the performance qualification test weldment is completed, the next step is to perform the required non-destructive examinations which are then followed by destructive testing. All of the testing can be performed in-house by qualified personnel using properly calibrated equipment and procedures, or the testing can be performed externally by a qualified testing laboratory. Further discussion of the examination and testing requirements and acceptance criteria for groove weld performance qualification are provided in

Sect. 8.4, for fillet weld performance qualification are provided in Sect. 8.5, for corrosion resistant overlay performance qualification in Sect. 8.6, and for hardfacing overlay performance qualification in Sect. 8.7.

### 8.1.4 Performance Qualification Documentation

Once the performance qualification test coupon has met all the examination and testing acceptance requirements, the welder/operator is now deemed qualified within the limits of the qualification and it's time to document the performance qualification. While there is no mandatory format, the performance qualification test record should identify the welding procedure specification (WPS) followed during the test, the required essential performance qualification variables used to produce the test coupon, the examination and testing methods along with their results, and the limits of qualification (ref. AWS B2.1 5.1.8 and ASME IX QW-104). Further discussion of the performance qualification record is provided in Sect. 3.3.

## 8.2 General Requirements

In addition to the considerations and requirements discussed above, the following are other general requirements to consider when qualifying multiple welding processes or multiple welders/operators in a single test weldment. Additional discussion is also provided for the simultaneous procedure and performance qualification, and simultaneous performance qualification of welders or operators by multiple organizations.

### 8.2.1 Multiple Welding Processes in a Single Test Weldment

Both AWS B2.1 and ASME IX allow multiple welding processes to be qualified on a single test coupon during performance qualification. The qualification thickness of deposited weld metal shall be considered individually for each welding process. A welder or welding operator qualified with multiple processes on a single test coupon is qualified to weld in production using any of these processes individually or in different combinations, provided they weld within the limits of their qualification with each specific process. Failure of any portion of the test coupon constitutes failure for all processes used in combination (ref. AWS B2.1 5.4.4 and ASME IX QW-306).

### 8.2.2 Multiple Welders/Operators in a Single Test Weldment

Both AWS B2.1 and ASME IX allow multiple welders and/or welding operators to qualify on a single test coupon during performance qualification using the same or a different welding process. The qualification thickness of deposited weld metal shall be considered individually for each welder/operator. A welder or welding operator qualified in combination on a single test coupon is qualified to weld in production using any of their qualified processes individually or in different combinations within the limits of qualification for each process. Failure of any portion of the test coupon constitutes failure for all welders/operators used in combination (ref. AWS B2.1 5.4.4 and ASME IX QW-306).

### 8.2.3 Simultaneous Procedure and Performance Qualification

Both AWS B2.1 and ASME IX allow simultaneous welding procedure qualification along with a welder or welding operator performance qualification on the same test weldment. A welder or welding operator who completes an acceptable welding procedure qualification test shall be qualified within the limits of the performance qualification variables tested (ref. AWS B2.1 5.1.12 and ASME IX QG-106.2(e), QW-301.2).

### 8.2.4 Simultaneous Performance Qualifications by Multiple Organizations

ASME IX allows multiple organizations to collectively perform performance qualification of one or more welder/operators simultaneously. Each participating organization must have a representative present to provide oversight of the simultaneous performance qualification. The provisions and limitations of the simultaneous performance qualification are provided in ASME IX QG-106.3. AWS B2.1 has no provisions for simultaneous performance qualifications by multiple organizations.

### 8.2.5 Performance Qualification Oversight and Organizational Responsibilities

Similar to the provisions outlined in Sect. 4.5.6 for procedure qualification, both AWS B2.1 and ASME IX require the welder/operator performance qualification process, including the welding of the qualification test weldment, be under full oversight by the qualifying organization (ref. AWS B2.1 5.1.7 and ASME IX QG-106). Personnel employed by the qualifying organization are responsible for supervision, control, evaluation, and acceptance of qualification testing. This oversight ensures compliance with the applicable qualification requirements. Some standards and/or contractual requirements may require independent oversight by an external 3rd party such as a Certified Welding Inspector or CWI (AWS), Authorized Inspector or AI (NB), or Examiner (ISO).

The qualifying organization is responsible for documenting the procedure qualification in the form of a Performance Qualification Test Record (PQTR), Welder Performance

Qualification (WPQ) record, or a Welder Operator Performance Qualification (WOPQ) record. The qualifying organization shall certify the qualification record meets all the applicable requirements. Retention of the test coupons and any documentation related to examination and testing is not required after the qualifying organization accepts and documents the results on the qualification record.

It is permitted to subcontract any or all the work necessary for preparing the materials to be joined, the subsequent work for preparing test specimens from the completed test joint, and the performance of nondestructive examination and testing, provided the qualifying organization accepts full responsibility for any subcontracted work. While AWS B2.1 does not address simultaneous performance qualification by multiple organizations, ASME IX has provisions which allow multiple organizations to qualify the performance of a welder/operator simultaneously (ref. ASME IX QG-106.3). In order to satisfy such provisions, each participating qualifying organization shall individually have full control and oversight of the qualification.

## 8.3 Welding Variables

The following tables provide the required essential welding variables for welder and welding operator performance qualification under AWS B2.1 5.6 and ASME IX QW-350 and QW-360. For convenience, the variables are grouped as discussed in Sect. 4.1.9. In addition, the variables are separated for manual and semiautomatic welding processes for welder performance qualification and machine/mechanized and automatic welding processes for welding operator performance qualification. The Essential (E) qualification variables are compared for the five most common arc welding procedures (SMAW, GMAW, FCAW, GTAW, SAW). Variables marked N/A are not addressed for a given standard. It should be noted that welder/operator performance qualification typically does not require supplementary essential variables nor non-essential variables. The tables should be regarded as illustrative only since the various standards are revised periodically.

### 8.3.1 Welding Process

See Tables 8.2 and 8.3.

### 8.3.2 Joint Design

See Tables 8.4 and 8.5.

### 8.3.3 Base Metal

See Tables 8.6 and 8.7.

### 8.3.4 Filler Metals

There are no filler metal variables for mechanized/machine and automatic welding operator performance qualification (Table 8.8).

### 8.3.5 Position

See Tables 8.9 and 8.10.

### 8.3.6 Preheat and Interpass Temperature

There are no preheat and interpass temperature variables for manual, semiautomatic, mechanized/machine, and automatic welder/welding operator performance qualification (ref. AWS B2.1 5.1.1 and ASME IX QW-301.2). Therefore, the preheat and interpass temperature requirements of the WPS being utilized for the performance qualification may be omitted.

### 8.3.7 Postweld Heat Treatment

There are no postweld heat treatment variables for manual, semiautomatic, mechanized/machine, and automatic welder/welding operator performance qualification (ref. AWS B2.1 5.1.1 and ASME IX QW-301.2). Therefore, the postweld heat treatment requirements of the WPS being utilized for the performance qualification may be omitted.

### 8.3.8 Shielding Gas

There are no shielding gas variables for mechanized/machine and automatic welding operator performance qualification (Table 8.11).

### 8.3.9 Electrical Characteristics

See Tables 8.12 and 8.13.

### 8.3.10 Other Variables

See Tables 8.14 and 8.15.

## 8.4 Groove Welder/Operator Qualification

The testing methods and the number of tests required for a groove welder/operator performance qualification test for both AWS B2.1 (ref. Tables 5.1 and 5.2) and ASME IX (ref. Table QW-452.1(a)) are summarized in Table 8.16.

Both AWS B2.1 and ASME IX allow a welder or welding operator who prepares a groove weld procedure qualification test coupon meeting all the requirements for procedure qualification to also be qualified within the limits of performance qualification (ref. AWS B2.1 5.1.12 and ASME IX QW-301.2).

Special test weldments shall be examined and tested as specified by the referencing document. When a test in Table 8.16 is specified by the referencing document, the acceptance criteria shall be as required in AWS B2.1 or ASME IX. The criteria for all other tests shall be as specified in the referencing document.

### 8.4.1 Groove Testing and Examination Requirements for AWS B2.1

*Visual Examination*

The test weld may be examined visually by the qualifier at any time, and the test terminated at any stage if the necessary skills are not exhibited. The completed test weld shall be visually examined without magnification (ref. AWS B2.1 5.5.1).

For standard test weldments greater or equal to 1/16 in. [1.5 mm] base metal thickness, the acceptance criteria for visual examination are as follows (ref. AWS B2.1 5.5.1.1(1)):

- No cracks or incomplete fusion.
- No incomplete joint penetration in groove welds, except where partial joint penetration groove welds are specified.
- Undercut depth shall not exceed the lesser of 10% of the base metal thickness or 1/32 in. [1 mm].
- Face reinforcement or root reinforcement shall not exceed 1/8 in. [3 mm].
- No single pore shall exceed 3/32 in. [2 mm] diameter.

**Table 8.2** Welding process variables for manual and semiautomatic welding processes

| AWS B2.1 | | | | | | | ASME IX | | | | | | |
|---|---|---|---|---|---|---|---|---|---|---|---|---|---|
| Variable | Ref. | SMAW | GMAW | FCAW | GTAW | SAW | Variable | Ref. | SMAW | GMAW | FCAW | GTAW | SAW |
| A change in welding process except that welders qualified with GMAW spray, pulsed spray, or globular transfer are also qualified to weld with gas shielded FCAW and vice versa | 5.6.1.1(a) | E | E | E | E | E | Each welder or welding operator shall be qualified within the limits given in QW-301 for the specific welding process(es) he will be required to use in production welding | QW-351 | E | E | E | E | E |
| | | | | | | | Semiautomatic GMAW also qualifies semiautomatic FCAW and vice-versa (ref. BPVC IX Interpretation IX-10-03 and X-86-46) | QW-355 | | E | | | |

**Table 8.3** Welding process variables for mechanized/machine and automatic welding processes

| AWS B2.1 | | | | | | | ASME IX | | | | | | |
|---|---|---|---|---|---|---|---|---|---|---|---|---|---|
| Variable | Ref. | SMAW | GMAW | FCAW | GTAW | SAW | Variable | Ref. | SMAW | GMAW | FCAW | GTAW | SAW |
| A change in welding process except that welding operators qualified with GMAW spray, pulsed spray, or globular transfer are also qualified to weld with gas shielded FCAW and vice versa | 5.6.1.2(a) | | E | E | E | E | A change in the welding process. (applies to machine and automatic welding) | QW-361.1(b) QW-362(a) | | E | E | E | E |

**Table 8.4** Joint design variables for manual and semiautomatic welding processes

| AWS B2.1 | | | | | | | ASME IX | | | | | | |
|---|---|---|---|---|---|---|---|---|---|---|---|---|---|
| Variable | Ref. | SMAW | GMAW | FCAW | GTAW | SAW | Variable | Ref. | SMAW | GMAW | FCAW | GTAW | SAW |
| The deletion of backing | 5.6.1.1(2) | E | E | E | E | E | The deletion of the backing in single welded groove welds. Double-welded groove welds are considered welding with backing | QW-402.4 | E | E | E | E | |

**Table 8.5** Joint design variables for mechanized/machine and automatic welding processes

| AWS B2.1 | | | | | | | ASME IX | | | | | | |
|---|---|---|---|---|---|---|---|---|---|---|---|---|---|
| Variable | Ref. | SMAW | GMAW | FCAW | GTAW | SAW | Variable | Ref. | SMAW | GMAW | FCAW | GTAW | SAW |
| The deletion of consumable inserts | 5.6.1.2(4) | | E | E | E | E | The deletion of consumable inserts, except that qualification with consumable inserts shall also qualify for fillet welds and welds with backing. (applies to *machine* welding only) | QW-361.2(f) | | | | E | |
| N/A | | | | | | | The deletion of backing. Double-welded groove welds are considered welding with backing. (applies to machine welding only) | QW-361.2(g) | | E | E | E | E |

**Table 8.6** Base metal variables for manual and semiautomatic welding processes

| AWS B2.1 | | | | | | | ASME IX | | | | | | |
|---|---|---|---|---|---|---|---|---|---|---|---|---|---|
| Variable | Ref. | SMAW | GMAW | FCAW | GTAW | SAW | Variable | Ref. | SMAW | GMAW | FCAW | GTAW | SAW |
| A change in base metal except as permitted in 5.4.1 | 5.6.1.1(4) | E | E | E | E | E | A change from one P-number to any other P-number or to a base metal not listed in Table QW/QB-422, except as permitted in QW-423, and in QW-420. For tube-to-tubesheet welding: a change in the P-number or A-number of the tubesheet cladding material (if the cladding material is part of the weld) | QW-403.18 | E | E | E | E | E |
| Qualification is valid only for metals having the same M-numbers, except as otherwise permitted in Table 5.3 | 5.4.1 | E | E | E | E | E | | | | | | | |
| N/A | | | | | | | A change in the pipe diameter beyond the range qualified in QW-452, except as otherwise permitted in QW-303.1, QW-303.2, QW-381.2(c), or QW-382.1(f). For tube-to-tubesheet welding: an increase or decrease greater than 10% of the specified tube diameter<br>(a) For a groove weld attaching a set-on nozzle or branch (with the weld preparation on the nozzle or branch), the range qualified from Table QW-452.3 shall be based on the nozzle or branch pipe O.D<br>(b) For a groove weld attaching a set-in nozzle or branch (with the weld preparation on the shell, head, or run pipe), the range qualified from Table QW-452.3 shall be based on the shell, head, or run pipe O.D | QW-403.16 | E | E | E | E | E |

**Table 8.7** Base metal variables for mechanized/machine and automatic welding processes

| AWS B2.1 | | | | | | | ASME IX | | | | | | |
|---|---|---|---|---|---|---|---|---|---|---|---|---|---|
| Variable | Ref. | SMAW | GMAW | FCAW | GTAW | SAW | Variable | Ref. | SMAW | GMAW | FCAW | GTAW | SAW |
| A change in base metal except as permitted in 5.4.1 | 5.6.1.1(4) | | E | E | E | E | N/A | | | | | | |
| Qualification is valid only for metals having the same M-numbers, except as otherwise permitted in Table 5.3 | 5.4.1 | | E | E | E | E | | | | | | | |

**Table 8.8** Filler metal variables for manual and semiautomatic welding processes

| AWS B2.1 | | | | | | | ASME IX | | | | | | |
|---|---|---|---|---|---|---|---|---|---|---|---|---|---|
| Variable | Ref. | SMAW | GMAW | FCAW | GTAW | SAW | Variable | Ref. | SMAW | GMAW | FCAW | GTAW | SAW |
| N/A | | | | | | | The deletion or addition of filler metal | QW-404.14 | | | | E | |
| A change in filler metal F-number except as allowed in 5.4.2 | 5.6.1.1(3) | E | E | E | E | E | A change from one F-number in Table QW-432 to any other F-number or to any other filler metal, except as permitted in QW-433 | QW-404.15 | E | E | E | E | E |
| Tests shall be performed using a filler metal which has an assigned F-number listed in Annex B. Table 5.4 provides a matrix showing filler metals which, if used in qualification testing, will qualify that welder and welding operator to use other filler metals without further testing. A test using a filler metal not assigned an F-number in Annex B shall qualify only for that filler metal | 5.4.2 | E | E | E | E | E | | | | | | | |
| For GMAW, GTAW, or PAW, omission or addition of consumable inserts | 5.6.1.1(10) | | E | | E | | The omission or addition of consumable inserts. Qualification in a single-welded butt joint, with or without consumable inserts, qualifies for fillet welds and single-welded butt joints with backing or double-welded butt joints. Consumable inserts that conform to SFA-5.30, except that the chemical analysis of the insert conforms to an analysis for any bare wire given in any SFA specification or AWS Classification, shall be considered as having the same F-number as that bare wire as given in Table QW-432 | QW-404.22 | | | | E | |

(continued)

**Table 8.8** (continued)

| AWS B2.1 | | | | | | | ASME IX | | | | | | |
|---|---|---|---|---|---|---|---|---|---|---|---|---|---|
| Variable | Ref. | SMAW | GMAW | FCAW | GTAW | SAW | Variable | Ref. | SMAW | GMAW | FCAW | GTAW | SAW |
| N/A | | | | | | | A change from one of the following filler metal product forms to another:<br>(a) bare (solid or metal cored)<br>(b) flux cored<br>(c) flux coated (solid or metal cored)<br>(d) powder | QW-404.23 | | | | E | |
| N/A | | | | | | | A change in deposited weld metal thickness beyond that qualified in accordance with QW-451 for procedure qualification or QW-452 for performance qualification, except as otherwise permitted in QW-303.1 and QW-303.2. When a welder is qualified using volumetric examination, the maximum thickness stated in Table QW-452.1(b) applies | QW-404.30 | E | E | E | E | E |
| N/A | | | | | | | For the low voltage short-circuiting type of gas metal-arc process when the deposited weld metal thickness is less than 1/2 in. (13 mm), an increase in deposited weld metal thickness beyond 1.1 times that of the qualification test deposited weld metal thickness. For weld metal thicknesses of 1/2 in. (13 mm) and greater, use Table QW-451.1, Table QW-451.2, or Tables QW-452.1(a) and QW-452.1(b), as applicable | QW-404.32 | | E | E | | |

**Table 8.9** Position variables for manual and semiautomatic welding processes

| AWS B2.1 | | | | | | | ASME IX | | | | | | |
|---|---|---|---|---|---|---|---|---|---|---|---|---|---|
| Variable | Ref. | SMAW | GMAW | FCAW | GTAW | SAW | Variable | Ref. | SMAW | GMAW | FCAW | GTAW | SAW |
| A change in position from that qualified, except as permitted in 5.4.3 and Table 5.5 | 5.6.1.1(7) | E | E | E | E | E | The addition of other welding positions than those already qualified. See QW-120, QW-130, QW-203, and QW-303 | QW-405.1 | E | E | E | E | E |
| Test coupons welded in the specific test positions described in Annex A1.1 and A1.3 qualify the welder to weld plate or pipe as permitted in Table 5.5. Weldment orientation other than the specific test positions shown in Annex A is permitted, but such tests qualify only for the orientation tested. Figures A.1B and A.1D show the permitted angular deviation in weld axis inclination and weld face rotation for each test position passed | 5.4.3 | | | | | | | | | | | | |
| A change in vertical weld progression from uphill to downhill, or vice versa for any pass except root passes that are completely removed by back gouging or final passes used to dress the final weld surface | 5.6.1.1(8) | E | E | E | E | E | A change from upward to downward, or from downward to upward, in the progression specified for any pass of a vertical weld, except that the cover or wash pass may be up or down. The root pass may also be run either up or down when the root pass is removed to sound weld metal in the preparation for welding the second side | QW-405.3 | E | E | E | E | |

**Table 8.10** Position variables for mechanized/machine and automatic welding processes

| AWS B2.1 | | | | | | | ASME IX | | | | | | |
|---|---|---|---|---|---|---|---|---|---|---|---|---|---|
| Variable | Ref. | SMAW | GMAW | FCAW | GTAW | SAW | Variable | Ref. | SMAW | GMAW | FCAW | GTAW | SAW |
| A change in position except as permitted in Table 5.5 | 5.6.1.1(3) | | E | E | E | E | The addition of welding positions other than those already qualified (see QW-120, QW-130, and QW-303). (applies to machine welding only) | QW-361.2(e) | | E | E | E | E |

**Table 8.11** Shielding gas variables for manual and semiautomatic welding processes

| AWS B2.1 | | | | | | | ASME IX | | | | | | |
|---|---|---|---|---|---|---|---|---|---|---|---|---|---|
| Variable | Ref. | SMAW | GMAW | FCAW | GTAW | SAW | Variable | Ref. | SMAW | GMAW | FCAW | GTAW | SAW |
| For GMAW, GTAW, or PAW, deletion of root shielding gas except for double welded butt joints, partial joint penetration groove, and fillet welds | 5.6.1.1(10) | | E | | E | | The omission of backing gas except that requalification is not required when welding a single welded butt joint with a backing strip or a double-welded butt joint or a fillet weld. This exception does not apply to P-No. 51 through P-No. 53, P-No. 61 through P-No. 62, and P-No. 10I metals | QW-408.8 | | E | E | E | |

**Table 8.12** Electrical characteristics variables for manual and semiautomatic welding processes

| AWS B2.1 | | | | | | | ASME IX | | | | | | |
|---|---|---|---|---|---|---|---|---|---|---|---|---|---|
| Variable | Ref. | SMAW | GMAW | FCAW | GTAW | SAW | Variable | Ref. | SMAW | GMAW | FCAW | GTAW | SAW |
| For GTAW, a change from alternating to direct current or vice versa, or a change in polarity | 5.6.1.1(6) | | | | E | | A change from AC to DC, or vice versa; and in DC welding, a change from electrode negative (straight polarity) to electrode positive (reverse polarity), or vice versa. This variable does not apply to a WPS qualified for welding base metals that are assigned to P-No. 8, P-Nos. 21 through 26, and P-Nos. 41 *through* 49 | QW-409.4 | | | | E | |
| For GMAW, a change from spray transfer, globular transfer, or pulsed spray welding to short-circuiting transfer; or vice versa | 5.6.1.1(9) | | E | | | | A change from globular, spray or pulsed spray transfer welding to short-circuiting transfer welding or vice versa | QW-409.2 | | E | E | | |

**Table 8.13** Electrical characteristics variables for mechanized/machine and automatic welding processes

| AWS B2.1 | | | | | | | ASME IX | | | | | | |
|---|---|---|---|---|---|---|---|---|---|---|---|---|---|
| Variable | Ref. | SMAW | GMAW | FCAW | GTAW | SAW | Variable | Ref. | SMAW | GMAW | FCAW | GTAW | SAW |
| For GMAW welding a change from any transfer mode to the short-circuiting mode | 5.6.1.2(5) | | E | | | | N/A | | | | | | |

**Table 8.14** Other variables for manual and semiautomatic welding processes

| AWS B2.1 | | | | | | | ASME IX | | | | | | |
|---|---|---|---|---|---|---|---|---|---|---|---|---|---|
| Variable | Ref. | SMAW | GMAW | FCAW | GTAW | SAW | Variable | Ref. | SMAW | GMAW | FCAW | GTAW | SAW |
| A change in thickness or diameter from that tested except as permitted in Tables 5.6, 5.7 and 5.8. Circumferential or radial fillet or groove weldments other than pipe (such as round stock or reinforcing plates), shall be considered the same as pipe, in accordance with Tables 5.5, 5.6, 5.7 and 5.8 | 5.6.1.1(11) | E | E | E | E | E | N/A | | | | | | |

**Table 8.15** Other variables for mechanized/machine and automatic welding processes

| AWS B2.1 | | | | | | | ASME IX | | | | | | |
|---|---|---|---|---|---|---|---|---|---|---|---|---|---|
| Variable | Ref. | SMAW | GMAW | FCAW | GTAW | SAW | Variable | Ref. | SMAW | GMAW | FCAW | GTAW | SAW |
| A change from direct visual to remote visual control or vice versa | 5.6.1.2(2) | | E | E | E | E | A change from direct visual control to remote visual control and vice-versa. (applies to machine welding only) | QW-361.2(b) | | E | E | E | E |
| N/A | | | | | | | A change from automatic to machine welding. (applies to automatic welding only) | QW-361.1(a) | | E | E | E | E |
| N/A | | | | | | | The deletion of an automatic arc voltage control system for GTAW. (applies to machine welding only) | QW-361.1(c) | | | | E | |
| N/A | | | | | | | The deletion of automatic joint tracking. (applies to machine welding only) | QW-361.1(d) | | E | E | E | E |
| N/A | | | | | | | A change from single pass per side to multiple passes per side but not the reverse. (applies to machine welding only) | QW-361.1(h) | | E | E | E | E |

**Table 8.16** Test methods required for groove weld performance qualification

| | AWS B2.1 | ASME IX |
|---|---|---|
| Visual examination | Yes | Yes |
| Bend testing | Yes | Yes |
| Volumetric examination | In lieu of bend testing (RT only) | In lieu of bend testing (RT or UT) |

For standard test weldments less than 1/16 in. [1.5 mm] base metal thickness, the acceptance criteria for visual examination are as follows (ref. AWS B2.1 5.5.1.1(3)):

- No cracks or incomplete fusion.
- No melt-through (burn-through) which results in a hole.
- No weld reinforcement for groove welds that exceed 1/8 in. [3 mm].
- No visible porosity or inclusions.

For workmanship test weldments, if visual examination is the only criterion for acceptance, all weld beads are subject to examination. The acceptance criteria for visual examination of workmanship test weldments are as follows (ref. AWS B2.1 5.5.1.2):

- No cracks or incomplete fusion.
- Undercut shall not exceed the lesser of 10% of the base metal thickness or 1/32 in. [1 mm] Except that 1/16 in. [1.5 mm] undercut is acceptable if less than 1/4 in. [6 mm] long, provided the accumulated length of such undercut is less than 3/4 in. [19 mm] in any 12 in. [305 mm] of weld.
- Reinforcement shall not exceed 1/8 in. [3 mm].
- No single pore shall exceed 3/32 in. [2 mm] diameter.

*Bend Testing*

The required number of bend tests depends on the test weldment form and test position and are summarized in Table 8.17 (ref. AWS B2.1 Table 5.2).

The bend test specimen blanks shall be removed from the locations shown in AWS B2.1 Figs. 5.3 or 5.4 for pipe, Fig. 5.5 for box tube, and Figs. 5.6 or 5.7 for plate. For base metal 3/8 in. [10 mm] upto 3/4 in. [19 mm] thick, side bends may be taken in lieu of face and root bends. For thicknesses over 3/4 in. [19 mm], side bends shall be used. Longitudinal bend specimens may be substituted for transverse bend specimens for welds that differ in bending behavior between two base metals or between base metal and weld metal.

The preparation and dimensions of specimen blanks for bend testing are provided in AWS B2.1 Annex A Figure A.2C for transverse side bend specimens, Figure A.2A for transverse face and root bend specimens, and Figure A.2B for longitudinal face and root

**Table 8.17** AWS B2.1 number of bend tests

| Position | Test weldment form | | | |
|---|---|---|---|---|
| | Plate | Pipe | Box tube | Sheet |
| 1 | qty. 2 | qty. 2 | qty. 2 | qty. 2 |
| 2 | qty. 2 | qty. 2 | qty. 2 | qty. 2 |
| 3 | qty. 2 | | | qty. 2 |
| 4 | qty. 2 | | | qty. 2 |
| 5 | | qty. 4 | qty. 4 | |
| 6 | | qty. 4 | qty. 4 | |

*Notes*
(1) Both AWS and ASME appends position with the letter "G" to denote a groove weld
(2) Position of welding:
1 = Flat
2 = Horizontal
3 = Vertical
4 = Overhead
5 = Horizontal pipe
6 = 45° Inclined pipe

bend specimens. The cut surfaces of Figures A.2A and A.2B are designated the specimen sides and the other two surfaces are designated the face and root surfaces. Subsize bend tests can be utilized for pipe 4 in. [102 mm] outside diameter or less (ref. AWS B2.1 A2.3). Nonstandard bend specimens can be utilized for base metal thicknesses less than 3/8 in. [10 mm] (ref. AWS B2.1 A2.4).

Weld reinforcement and backing of face- and root- bend specimens shall be removed flush with the specimen surface. Cut surfaces shall be parallel, may be thermally cut, and shall be machined or ground a minimum of 1/8 in. [3 mm] on thermally cut edges, except that M-1 metals may be bent "as-cut" if no objectionable surface roughness exists (ref. AWS B2.1 A2.2).

Bend specimens shall be bent in one of the guided bend test fixtures shown in AWS B2.1 Annex A Figure A.5A bottom ejecting guided bend fixture, Figure A.5B bottom type guided bend fixture, or Figure A.5C wrap around guided bend fixture. For face bend specimens, the weld face side shall be on the convex side of the bend specimen. For root bend specimens, the weld root side shall be on the convex side of the bend specimen. Side bend specimens may be bent in either direction. For all transverse bend specimens, the weld metal and heat-affected zone shall be completely within the bent portion of the specimen after bending (ref. AWS B2.1 5.5.3.1).

The acceptance criteria for bend testing in AWS B2.1 5.5.3.2 states:

- There shall be no open discontinuity exceeding 1/8 in. [3 mm], measured in any direction on the convex surface of the specimen after bending.
- Any cracks occurring on the corners of the specimen during bending shall not be considered, unless there is definite evidence that they result from slag inclusions or other discontinuities.

*Volumetric Examination*

AWS B2.1 allows substitution of radiographic examination instead of mechanical testing for groove weld performance qualification. Unless otherwise specified in the referencing document, the radiographic procedure and acceptance criteria shall be in accordance with AWS B2.1 Annex D as follows (ref. AWS B2.1 5.5.2). Radiography may be substituted for bend testing for the SMAW, GTAW, GMAW (except short-circuiting), FCAW, and SAW processes, as applicable, for qualifications on all base metals except M-51 to M-55 and M-61 to M-62. GTAW tests in M-51 to M-55 and M-61 to M-62 may be qualified with radiography (ref. AWS B2.1 Table 5.1 note a). Unless otherwise specified in the referencing document, the radiographic procedure and acceptance criteria shall be as follows.

The entire weld, except for the discard on plate, shall be examined. When the qualification test is a Standard Test Weldment, a minimum of 6 in. [152 mm] of weld length shall be examined, except that for pipe the entire weld shall be examined. Multiple welds may be required for small diameter test weldments to permit an examination of 6 in. [152 mm] of weld. No more than four joints need be examined to meet this requirement. For welder operators, the minimum length of weld examined shall be 3 ft. [1 m]. Face reinforcement may be removed at the option of the Qualifier. Root reinforcement or backing strips shall not be removed from single-welded groove joints. The backing width shall be 3 in. [76 mm] minimum (ref. AWS B2.1 Fig. 5.6 Note 4).

When qualification is based upon a production weldment the criteria for acceptance of the weld shall be as required by the referencing document, otherwise the acceptance shall be in accordance with the following requirements (ref. AWS B2.1 D4.4):

- There shall be no cracks, incomplete joint penetration, or incomplete fusion. Root concavity in the test weldment is permitted, provided the film density through the weld is not less than that through the base metal.
- Acceptable linear indications shall be as shown in AWS B2.1 Table D.1. Linear discontinuities are those in which the length is more than three times the width.
- Acceptable rounded indications shall be as shown in AWS B2.1 Table D.2 and Figure D.1. Rounded discontinuities are those having a length less than three times the width and may be circular, elliptical, or irregular in shape.

### 8.4.2 Groove Qualification Limits for AWS B2.1

In addition to the restrictions of specific variables for the different welding processes as outlined in Table 8.2 through Table 8.15, additional limitations for position, base metal, filler metal, base metal thickness, pipe diameter, and weld deposit thickness are discussed below for groove welder/operator performance qualification for AWS B2.1.

*Position*

The standard welding positions for groove welds in plate or pipe are defined in AWS B2.1 Figure A.1A. AWS B2.1 defines a change in position as an angular deviation of>±15 degrees from the specified horizontal and vertical planes, and an angular deviation of>±5 degrees from the specified inclined plane are permitted during welding. This is depicted for groove welds in AWS B2.1 Figure A.1B.

Test coupons welded in the specific test positions described in AWS B2.1 Annex A1.1 qualify the welder to weld plate or pipe as permitted in Table 8.18 (ref. AWS B2.1 Table 5.5). Qualification on a complete joint penetration groove weld also qualifies the welder or welding operator for partial joint penetration groove welds, fillet welds and tack welds (ref. AWS B2.1 5.1.14). Welders qualified on tubular product forms may weld on both tubular and plate in accordance with any restrictions on diameter (ref. AWS B2.1 Table 5.5 note a). Weldment orientation other than the specific test positions shown in Annex A is permitted, but such tests qualify only for the orientation tested (ref. B2.1 5.4.3).

#### 8.4.2.1 Base Metal

Welder/operator performance qualification is valid only for materials having the same M-numbers as utilized in the test weldment except as otherwise permitted in Table 8.19 (ref. AWS B2.1 5.4.1 and Table 5.3). If materials not listed in AWS B2.1 Annex C are used for qualification tests, the welder or welding operator shall be qualified to weld only on the material used in the test weldment. In addition to being qualified for the materials listed in Table 8.19, the welder/operator is also qualified for unlisted materials of similar chemical composition to the test materials.

*Filler Metal*

Tests shall be performed using a filler metal which has an assigned F-Number listed in AWS B2.1 Annex B. Table 8.20 (ref. AWS B2.1 5.4.2 and Table 5.5) provides a matrix showing filler metals which, if used in qualification testing, will qualify that welder and welding operator to use other filler metals without further testing. For a test using F-Number 6 filler metal, deposited solid bare wire which is not covered by an AWS specification but which conforms to an A-Number analysis in AWS B2.1 Annex B Table B.2, may be considered classified as F-Number 6. A test using a filler metal not assigned an F-Number in AWS B2.1 Annex B shall qualify only for that filler metal.

**Table 8.18** AWS B2.1 qualified positions for groove and fillet welds

| Position(s) tested | | Position(s) qualified | | |
|---|---|---|---|---|
| | | Groove in plate and pipe > 24 in. [610 mm] O.D | Groove in pipe ≤24 in. [610 mm] O.D | Fillet in plate and pipe |
| Plate | 1 | 1 | | 1, 2 |
| | 2 | 1, 2 | | 1, 2 |
| | 3 | 1, 3 | | 1, 2, 3 |
| | 4 | 1, 4 | | 1, 2, 4 |
| | 3 + 4 | 1, 3, 4 | | 1, 2, 3, 4 |
| | 2 + 3 + 4 | 1, 2, 3, 4 | | 1, 2, 3, 4 |
| Pipe | 1 | 1 | 1 | 1, 2 |
| | 2 | 1, 2 | 1, 2 | 1, 2 |
| | 5 | 1, 3, 4 | 1, 3, 4 | 1, 2, 3, 4,5, 6 |
| | 6 | 1, 2, 3, 4, 5, 6 | 1, 2, 3, 4, 5, 6 | 1, 2, 3, 4, 5, 6 |
| | 2 + 5 | 1, 2, 3, 4, 5, 6 | 1, 2, 3, 4, 5, 6 | 1, 2, 3, 4, 5, 6 |

*Notes*
(1) AWS appends position with the letter "G" to denote groove weld
(2) Position of welding:
1 = Flat
2 = Horizontal
3 = Vertical
4 = Overhead
5 = Horizontal pipe
6 = 45° Inclined pipe

**Table 8.19** AWS B2.1 base metal qualification ranges

| Test weldment material | Qualified materials |
|---|---|
| M-1 through M-11, M-34, and M-41 through M-47 | M-1 through M-11, and M-41 through M-47 and M-34 |
| M-21 through M-27 | Any M-21 through M-27 material |
| M-31 through M-33 and M-35 | Only the specific M-number used in the qualification test |
| M-34 or M-42 | Any M-34 and M-41 through M-47 material |
| M-51 through M-54 and M-61 and M62 | M-51 through M-54 and M-61 and M-62 |
| M-81 or M-83 | Any M-81 and M-83 |

**Table 8.20** AWS B2.1 filler metal qualification ranges

| Filler metal used for qualification | Filler metals qualified |
|---|---|
| F-number 1 through 5 | F-number used in the test and any lower F-number |
| F-number 6 | All F-number 6 filler metals |
| F-number 2X | All F-number 2X filler metals |
| F-number 3X | Only for the specific F-number 3X filler metal |
| F-number 4X | F-number 1 through 5 and all F-number 4X |
| F-number 5X | All F-number 5X filler metals |
| F-number 61 | All F-number 61 filler metals |
| F-number 71 | Only for the specific F-number 71 filler metal |
| F-number 91 | All F-number 91 filler metals |

*Base Metal Thickness*
Base metal thickness is not restricted for welder/operator groove weld performance qualification. The welder or welding operator is only restricted by weld deposit thickness (see below).

*Weld Deposit Thickness*
The welder/operator qualified groove weld deposit thickness for plate, pipe, and tube are shown in Table 8.21 (ref. AWS B2.1 Tables 5.6 and 5.7). Circumferential or radial groove weldments other than pipe (such as round stock or reinforcing plates), shall be considered the same as pipe (ref. AWS B2.1 5.6.1.1(11)). For GMAW-S, the maximum weld metal thickness deposited shall not exceed 1.1 times the thickness of weld metal deposited by the GMAW-S process in the performance qualification test.

*Pipe Diameter*
The welder/operator qualified groove diameter for pipe and tube are shown in Table 8.22 (ref. AWS B2.1 Table 5.6). Circumferential or radial groove weldments other than pipe (such as round stock or reinforcing plates) shall be considered the same as pipe (ref. AWS B2.1 5.6.1.1(11)).

**Table 8.21** AWS B2.1 qualified groove weld deposit thickness for plate, pipe, and tube

| Test weldment deposited weld metal thickness (t) in. [mm] | Weld metal deposit thickness qualified |
|---|---|
| | Maximum in. [mm] |
| <3/4 [19] | 2t |
| ≥3/4 [19] | Unlimited |

**Table 8.22** AWS B2.1 qualified groove weld diameter for pipe and tube

| Test weldment outside diameter in. [mm] | Outside diameter qualified |
|---|---|
| | Minimum in. [mm] |
| <1 [25] | Size welded |
| 1 through 2–7/8 [25–73] | 1 [25] |
| >2–7/8 [73] | >2–7/8 [73] |

### 8.4.3 Groove Testing and Examination Requirements for ASME IX

*Visual Examination*

For plate coupons all surfaces (except areas designated "discard") shall be examined visually before the cutting of bend specimens. Pipe coupons shall be visually examined over the entire circumference, inside and outside (ref. ASME IX QW-302.4). The acceptance criteria for visual examination of performance test coupons are as follows (ref. ASME IX QW-194):

- No cracks
- Complete joint penetration with complete fusion of weld metal and base metal.

*Bend Testing*

The required number and type of bend tests depends on the weld deposit thickness of the test weldment and are summarized in Table 8.23 (ref. ASME IX Table QW-452.1(a)). Note the "Thickness of Weld Deposit" is the total weld metal thickness deposited by all welders and all processes in the test coupon exclusive of the weld reinforcement.

The bend test specimen blanks shall be removed from locations shown in Figures QW-463.1(d), (e), and (f) for pipe, and Figures QW-463.1(a), (b), and (c) for plate. One face and root bend may be substituted for the two side bends for weld deposit thicknesses greater than 3/8 in. [10 mm]. When qualifying in positions 5G or 6G, a total of four bend specimens are required, but two face and two root bends may be substituted for the four side bends in accordance with Figure QW-463.2(d). When qualifying using a combination of 2G and 5G in a single test coupon, a total of six bend specimens are required, but three

**Table 8.23** ASME IX type and number of bend test specimens

| Thickness of weld deposit (t) in. [mm] | Side bend | Face bend | Root bend |
|---|---|---|---|
| t < 3/8 [10] | | 1 | 1 |
| 3/8 [10] ≥ t < 3/4 [19] | 2 | | |
| t ≥ 3/4 [19] | 2 | | |

face and three root bends may be substituted for the six side bends in accordance with Figure QW-463.2(f) or Figure QW-463.2(g). Coupons tested by face and root bends shall be limited to weld deposit made by one welder with one or two processes or two welders with one process each. Weld deposit by each welder and each process shall be present on the convex surface of the appropriate bent specimen.

The preparation and dimensions of specimen blanks for bend testing are provided in ASME IX QW-161.1 and Figure QW-462.2 for transverse side bend specimens, QW-161.2, QW-161.3 and Figure QW-462.3(a) for transverse face and root bend specimens, and QW-161.6, QW161.7 and Figure QW-462.3(b) for longitudinal face and root bend specimens. Dimensions for subsize transverse face and root bends are given in ASME IX QW-161.4 and Figure QW-462.3(a) general note (b). Weld reinforcements and backing strip or backing ring, if any, shall be removed essentially flush with the undisturbed surface of the base material.

Bend specimens shall be bent in one of the guided bend test fixtures shown in Figure QW-466.1 test jig dimensions, Figure QW-466.2 guided-bend roller jig, or Figure QW-466.3 guided-bend wrap around jig. The weld and heat-affected zone of a transverse weld bend specimen shall be completely within the bent portion of the specimen after testing.

The acceptance criteria for bend testing in QW-163 states:

- The guided-bend specimens shall have no open discontinuity in the weld or heat-affected zone exceeding 1/8 in. [3 mm], measured in any direction on the convex surface of the specimen after bending.
- Open discontinuities occurring on the corners of the specimen during testing shall not be considered unless there is definite evidence that they result from lack of fusion, slag inclusions, or other internal discontinuities.

*Volumetric Examination*

ASME IX allows the substitution of Radiographic or Ultrasonic examination instead of mechanical testing for groove weld performance qualification (ref. ASME IX QW-143). Welders and welding operators may be qualified by volumetric NDE when making a groove weld using SMAW, SAW, GTAW, GMAW (except short-circuiting mode for radiographic examination) or a combination of these processes, except for P-No. 21 through P-No. 26, P-No. 51 through P-No. 53, and P-No. 61 through P-No. 62 metals. Welders and welding operators making groove welds in P-No. 21 through P-No. 26, P-No. 51 through P-No. 53, P-No. 61, and P-No. 62 metals with the GTAW process may also be qualified by volumetric NDE per QW-191 (ref. ASME IX QW-304 and QW-305).

For standard test weldments the minimum length of coupon(s) to be examined shall be 6 in. (150 mm) and shall include the entire weld circumference for pipe(s), except that for small diameter pipe, multiple coupons of the same diameter pipe may be required, but the number need not exceed four consecutively made test coupons (ref. ASME IX QW-302.2). For production test weldments a minimum 6 in. (150 mm) length of the first production weld(s) made by a welder/operator using the process(es) and/or mode of arc transfer specified in QW-304 may be examined by volumetric NDE (ref. ASME IX QW-304.1 and QW-305.1). For pipe(s) welded in the 5G, 6G, or special positions, the entire production weld circumference made by the welder shall be examined. For small diameter pipe where the required minimum length of weld cannot be obtained from a single production pipe circumference, additional consecutive circumferences of the same pipe diameter made by the welder shall be examined, except that the total number of circumferences need not exceed four. Weld reinforcement may be removed or left in place but shall not be considered when determining the thickness for which the welder is qualified (ref. ASME IX QW-191.1.2.2).

For radiographic examination, welds in test assemblies and production weldments shall be judged unacceptable when the radiograph exhibits any imperfections in excess of the limits specified below (ref. ASME IX QW-191.1.2.2):

- Linear Indications
  - any type of crack or zone of incomplete fusion or penetration
  - any elongated slag inclusion which has a length greater than (−a) 1/8 in. [3 mm] for thicknesses up to 3/8 in. [10 mm], inclusive (−b) 1/3 the thickness when the thickness is greater than 3/8 in. [10 mm] to 2 1/4 in. [57 mm], inclusive (−c) 3/4 in. [19 mm] for thicknesses greater than 2 1/4 in. [57 mm]
  - any group of slag inclusions in line that have an aggregate length greater than the thickness in a length of 12 times that thickness, except when the distance between the successive imperfections exceeds 6L where L is the length of the longest imperfection in the group.
- Rounded Indications
  - The maximum permissible dimension for rounded indications shall be 20% of the thickness or 1/8 in. [3 mm], whichever is smaller.
  - For welds in material less than 1/8 in. [3 mm] in thickness, the maximum number of acceptable rounded indications shall not exceed 12 in a 6 in. [150 mm] length of weld. A proportionately fewer number of rounded indications shall be permitted in welds less than 6 in. [150 mm] in length.
  - For welds in material 1/8 in. [3 mm] or greater in thickness, the charts in Figure QW-191.1.2.2(b)(4) represent the maximum acceptable types of rounded indications illustrated in typically clustered, assorted, and randomly dispersed configurations. Rounded indications less than 1/32 in. [0.8 mm] in maximum diameter shall not be considered in the radiographic acceptance tests of welders and welding operators in these ranges of material thicknesses.

For ultrasonic examination, welds in test assemblies and production weldments shall be judged unacceptable when indications are evaluated as follows (ref. ASME IX QW-191.2.2):

- All indications characterized as cracks, lack of fusion, or incomplete penetration are unacceptable regardless of length.
- Indications exceeding 1/8 in. [3 mm] in length are considered relevant, and are unacceptable when their lengths exceed:
  - 1/8 in. [3 mm] for thicknesses up to 3/8 in. [10 mm], inclusive
  - 1/3 the thickness for thicknesses greater than 3/8 in. up to 2 1/4 in. [10 mm to 57 mm], inclusive
  - 3/4 in. [19 mm] for thicknesses greater than 2 1/4 in. [57 mm].

### 8.4.4 Groove Qualification Limits for ASME IX

In addition to the restrictions of specific variables for the different welding processes as outlined in Table 8.2 through Table 8.15, additional limitations for position, base metal, filler metal, base metal thickness, pipe diameter, and weld deposit thickness are discussed below for groove welder/operator performance qualification for ASME IX.

*Position*

The standard welding positions for groove welds in plate or pipe are defined in ASME IX Figures QW-461.3 and QW-461.4. ASME IX define a change in position as an angular deviation of>±15 degrees from the specified horizontal and vertical planes, and an angular deviation of>±5 degrees from the specified inclined plane are permitted during welding. This is depicted for groove welds in ASME IX QW-461.1.

Welders and welding operators who pass the required tests for groove welds in the test positions shown in Table 8.24 (ref. ASME IX Table QW-461.9) shall be qualified for the positions of groove welds, tack welds in joints to be groove or fillet welded, and fillet welds shown in Table 8.24. In addition, welders and welding operators who pass the required tests for groove welds shall also be qualified to make fillet welds in all thicknesses and pipe diameters of any size within the limits of the welding variables listed in Table 8.2 through Table 8.15, and tack welds in joints to be groove or fillet welded as limited in Table 8.24, as applicable. Note tack welds are not limited by pipe or tube diameters when their aggregate length does not exceed 25% of the weld circumference.

An organization who does production welding in a special orientation may make the tests for performance qualification in this specific orientation. Such qualifications are valid only for the flat position and for the special positions actually tested, except that an

**Table 8.24** ASME IX qualified positions for groove welds

| Position(s) tested | | Position(s) qualified | | |
|---|---|---|---|---|
| | | Groove in plate and pipe > 24 in. [610 mm] O.D | Groove in pipe ≤24 in. [610 mm] O.D | Fillet in plate and pipe |
| Plate | 1 | 1 | 1 | 1, 2 |
| | 2 | 1, 2 | 1, 2 | 1, 2 |
| | 3 | 1, 3 | 1 | 1, 2, 3 |
| | 4 | 1, 4 | 1 | 1, 2, 4 |
| | 3 + 4 | 1, 3, 4 | 1 | 1, 2, 3, 4 |
| | 2 + 3 + 4 | 1, 2, 3, 4 | 1, 2 | 1, 2, 3, 4 |
| | SP | 1, SP | 1, SP | 1, SP |
| Pipe | 1 | 1 | 1 | 1 |
| | 2 | 1, 2 | 1, 2 | 1, 2 |
| | 5 | 1, 3, 4 | 1, 3, 4 | 1, 2, 3, 4, 5, 6 |
| | 6 | 1, 2, 3, 4,5, 6 | 1, 2, 3, 4, 5, 6 | 1, 2, 3, 4, 5, 6 |
| | 2 + 5 | 1, 2, 3, 4, 5, 6 | 1, 2, 3, 4, 5, 6 | 1, 2, 3, 4, 5, 6 |
| | SP | 1, SP | 1,SP | 1, SP |

*Notes*
(1) ASME appends position with the letter "G" to denote groove weld
(2) Position of welding:
1 = Flat
2 = Horizontal
3 = Vertical
4 = Overhead
5 = Horizontal pipe
6 = 45° Inclined pipe
SP = Special positions

angular deviation of ±15 deg is permitted in the inclination of the weld axis and the rotation of the weld face, as defined in IX Figures QW-461.1 and QW-461.2.

*Base Metal*
Welder/operator performance qualification is valid only for materials having the same P-numbers as utilized in the test weldment except as otherwise permitted in Table 8.25 (ref. ASME IX QW-403.18 and QW-423). If materials not listed in ASME IX Table QW-422 are used for qualification tests, the welder or welding operator shall be qualified to weld only on the material used in the test weldment within the provisions of Table 8.25. Base metal used for welder qualification may be substituted for the base metal specified in the WPS in accordance with the following table. Any base metal shown in the same

**Table 8.25** ASME IX base metal qualification ranges

| Test weldment material | Qualified materials |
|---|---|
| P-No. 1 through P-No. 15F, P-No. 34, and P-No. 41 through P-No. 49 | P-No. 1 through P-No. 15F, P-No. 34, and P-No. 41 through P-No. 49 |
| P-No. 21 through P-No. 26 | Any P-No. 21 through P-No. 26 |
| P-No. 51 through P-No. 53 or P-No. 61 or P-No. 62 | P-No. 51 through P-No. 53 and P-No. 61 and P-No. 62 |
| Any unassigned metal to the same unassigned metal | The unassigned metal to itself |
| Any unassigned metal to any P-number metal | The unassigned metal to any metal assigned to the same P-number as the qualified metal |
| Any unassigned metal to any other unassigned metal | The first unassigned metal to the second unassigned metal |

row may be substituted in the performance qualification test coupon for the base metal(s) specified in the WPS followed during welder qualification. When a base metal shown in the left column of the table is used for welder qualification, the welder is qualified to weld all combinations of base metals shown in the right column, including unassigned metals of similar chemical composition to these metals (ref. ASME IX QW-423.1). A base metal used for welder qualification conforming to national or international standards or specifications may be considered as having the same P-Number as an assigned metal provided it meets the mechanical and chemical requirements of the assigned metal (ref. ASME IX QW-423.2).

*Filler Metal*

Welder/operator performance qualification is valid only for filler metals having the same F-Numbers as utilized in the test weldment except as otherwise permitted in Table 8.26 (ref. ASME IX QW-403.18 and QW-433). The F-Number grouping of electrodes and welding rods in are outlined in ASME IX Table QW-432. ASME IX does not have provisions for qualifying with un-assigned filler metals.

*Base Metal Thickness*

Base metal thickness is not restricted for welder/operator groove weld performance qualification. The welder or welding operator is only restricted by weld deposit thickness (see below).

*Weld Deposit Thickness*

The welder/operator qualified groove weld deposit thickness for plate, pipe, and tube are shown in Table 8.27 (ref. ASME IX Table QW-452.1(b)). When more than one welder, process, or set of essential variables is used during welding of a test coupon, the thickness,

**Table 8.26** ASME IX filler metal qualification ranges

| Filler metal used for qualification | Filler metals qualified |
|---|---|
| F-No. 1 with backing | F-No. 1 with backing |
| F-No. 1 without backing | F-No. 1 with or without backing |
| F-No. 2 with backing | F-No. 1 or F-No. 2 with backing |
| F-No. 2 without backing | F-No. 1 with backing, F-No.2 with or without backing |
| F-No. 3 with backing | F-No. 1, F-No. 2, or F-No. 3 with backing |
| F-No. 3 without backing | F-No. 1 or F-No.2 with backing, F-No.3 with or without backing |
| F-No. 4 with backing | F-No. 1, F-No. 2, F-No.3, or F-No. 3 with backing |
| F-No. 4 without backing | F-No. 1, F-No.2, or F-No.3 with backing, F-No.4 with or without backing |
| F-No. 5 with backing | F-No. 1 or F-No. 5 with backing |
| F-No. 5 without backing | F-No. 1 with backing, F-No.5 with or without backing |
| F-No. 6 | All F-No. 6 filler metals[1] |
| F-No. 21 through F-No. 26 | All F-No. 21 through F-No. 26 |
| F-No. 31 through F-No. 33<br>F-No. 35 through F-No. 37 | Only for the same F-No. as was used during the qualification test |
| F-No. 34 or F-No.41 though F-No.46 | F-No. 34 and all F-No.41 though F-No.46 |
| F-No. 51 through F-No.55 | All F-No. 51 through F-No. 55 |
| F-No. 61 | All F-No. 61 |
| F-No. 71 or F-No.72 | Only for the same F-No. as was used during the qualification test |

*Note*

(1) Deposited weld metal made using a bare rod not covered by an SFA Specification but which conforms to an analysis listed in Table QW-442 shall be considered to be classified as F-No. 6

*t*, of the weld metal in the coupon deposited by each welder, for each process, and with each set of essential variables shall be determined and used individually in the "Test Weldment Deposited Weld Metal Thickness (t)" column to determine the "Weld Metal Deposit Thickness Qualified". Two or more pipe test coupons with different weld metal thickness may be used to determine the weld metal thickness qualified and that thickness may be applied to production welds using the smallest diameter for which the welder is qualified in accordance with pipe diameter thickness limitations of the next section below.

#### 8.4.4.1 Pipe Diameter Thickness

The welder/operator qualified groove diameter for pipe, and tube are shown in Table 8.28 (ref. ASME IX Table QW-452.3).

**Table 8.27** ASME IX qualified groove weld deposit thickness for plate, pipe, and tube

| Test weldment deposited weld metal thickness ($t$) in. [mm] | Weld metal deposit thickness qualified |
|---|---|
| | Maximum in. [mm] |
| <1/2 [13] | 2t |
| ≥1/2 [13] with < 3 layers | 2t |
| ≥1/2 [13] with ≥ 3 layers | Unlimited |

**Table 8.28** ASME IX qualified groove weld diameter for pipe and tube

| Test weldment outside diameter in. [mm] | Outside diameter qualified |
|---|---|
| | Minimum in. [mm] |
| <1 [25] | Size welded |
| 1 through 2–7/8 [25–73] | 1 [25] |
| >2–7/8 [73] | >2–7/8 [73] |

## 8.5 Fillet Welder/Operator Qualification

The testing methods and the number of tests required for a fillet welder/operator performance qualification test for both AWS B2.1 (ref. Table 5.1) and ASME IX (ref. Table QW-452.5) are summarized in Table 8.29.

AWS B2.1 allows performance qualification on a complete penetration groove-weld test coupon to qualify for fillet welds, but not vice versa (ref. AWS B2.1 5.1.14). Additionally, AWS B2.1 allows a welder/operator who completes an acceptable fillet weld procedure qualification test to also be qualified within the limits for performance qualification for fillet welds (ref. AWS B2.1 5.1.12). ASME IX also allows a welder or welding operator who prepares a fillet weld procedure qualification test coupon that

**Table 8.29** Test methods required for fillet weld performance qualification

| | AWS B2.1 | ASME IX |
|---|---|---|
| Visual examination | Yes | Not required |
| Macro examination | Yes[a] (qty.2) | Yes (qty. 1) |
| Break testing | Yes[a,b] | Yes |
| Shear testing | Optional[b] | Not optional |

*Notes*

(a) Only required for pipe or plate ≥1/16 in. [1.5 mm] thick

(b) The requirement for a break test and macro is waived for welders who successfully complete fillet procedure qualification tests (where shear tests, plus visual and macro examinations are used) meets all the requirements for procedure qualification to be also qualified within the limits for performance qualification for non-pressure retaining fillet welds (ref. ASME IX QW-301.2).

Special test weldments shall be examined and tested as specified by the referencing document. When a test in Table 8.12 is specified by the referencing document, the acceptance criteria shall be as required in AWS B2.1 or ASME IX. The criteria for all other tests shall be as specified in the referencing document.

## 8.5.1 Fillet Testing and Examination Requirements for AWS B2.1

### 8.5.1.1 Visual Examination

The test weld may be examined visually by the Qualifier at any time, and the test terminated at any stage if the necessary skills are not exhibited. The completed test weld shall be visually examined without magnification (ref. AWS B2.1 5.5.1).

For standard test weldments greater or equal to than 1/16 in. [1.5 mm] base metal thickness, the acceptance criteria for visual examination are as follows (ref. AWS B2.1 5.5.1.1(1)):

- No cracks or incomplete fusion.
- No incomplete joint penetration in groove welds, except where partial joint penetration groove welds are specified.
- Undercut depth shall not exceed the lesser of 10% of the base metal thickness or 1/32 in. [1 mm].
- Face reinforcement or root reinforcement shall not exceed 1/8 in. [3 mm].
- No single pore shall exceed 3/32 in. [2 mm] diameter.
- For fillet weld tests, concavity or convexity of the weld face shall not exceed 1/16 in. [1.5 mm]. The two fillet weld sizes shall not differ by more than 1/8 in. [3 mm].

For standard test weldments less than 1/16 in. [1.5 mm] base metal thickness, the acceptance criteria for visual examination are as follows (ref. AWS B2.1 5.5.1.1(3)):

- No cracks or incomplete fusion.
- No melt-through (burn-through) which results in a hole.
- No convexity for fillet welds that exceed 1/8 in. [3 mm].
- No visible porosity or inclusions.

For workmanship test weldments, if visual examination is the only criterion for acceptance, all weld beads are subject to examination. The acceptance criteria for visual examination of workmanship test weldments are as follows (ref. AWS B2.1 5.5.1.2):

- No cracks or incomplete fusion.
- Undercut shall not exceed the lesser of 10% of the base metal thickness or 1/32 in. [1 mm] Except that 1/16 in. [1.5 mm] undercut is acceptable if less than 1/4 in. [6 mm] long, provided the accumulated length of such undercut is less than 3/4 in. [19 mm] in any 12 in. [305 mm] of weld.
- Reinforcement shall not exceed 1/8 in. [3 mm].
- No single pore shall exceed 3/32 in. [2 mm] diameter.

*Macro Examination*

Macro examination is only required for pipe or plate $\geq$1/16 in. [1.5 mm] thick. Specimens shall be polished and etched to provide a clear definition of the weld metal and heat-affected zone (see AWS B2.1 Annex G). Visual examination of etched surfaces shall be without magnification. Both weld cross sections of the macroetch specimen from the fillet weld shall be examined. The weld cross section shall show no incomplete fusion and no cracks. Discontinuities at the weld root, not exceeding 1/32 in. [1 mm], shall be acceptable (ref AWS B2.1 5.5.5).

*Break Testing*

Bend-break testing is only required for pipe or plate $\geq$1/16 in. [1.5 mm] thick. Specimens shall be bent with the weld root in tension until the specimen either fractures or until it is bent flat upon itself. The specimen shall be accepted if (ref. AWS B2.1 5.5.4):

- The specimen does not fracture; or
- If the fillet fractures, the fractured surface shall exhibit no cracks or incomplete root fusion and no inclusion or porosity in the fracture surface exceeding 3/32 in. [2 mm] in its greatest dimension; or
- The sum of the greatest dimension of all inclusions and porosity do not exceed 3/8 in. [10 mm] in the specimen length.

*Shear Testing*

See the requirements for shear testing for a fillet weld procedure qualification in Sect. 5.2.2.

### 8.5.2 Fillet Qualification Limits for AWS B2.1

In addition to the restrictions of specific variables for the different welding processes as outlined in Table 8.2 through Table 8.15, additional limitations for Position, Base Metal, Filler Metal, Base Metal Thickness, Pipe Diameter, and Weld Deposit Thickness are discussed below for fillet welder/operator performance qualification for AWS B2.1.

*Position*

The standard welding positions for fillet welds in plate or pipe are defined in AWS B2.1 Figure A.1C. AWS B2.1 defines a change in position as an angular deviation of>±15 degrees from the specified horizontal and vertical planes, and an angular deviation of>±5 degrees from the specified inclined plane are permitted during welding. This is depicted for fillet welds in AWS B2.1 Figure A.1D.

Test coupons welded in the specific test positions described in AWS B2.1 Annex A1.1 qualify the welder/operator to weld plate or pipe as permitted in Table 8.30 (ref. AWS B2.1 Table 5.5). Qualification on a fillet weld qualifies only for fillet welding and tack welding (ref. AWS B2.1 5.1.14). Weldment orientation other than the specific test positions shown in Annex A is permitted, but such tests qualify only for the orientation tested (ref. AWS B2.1 5.4.3).

*Base Metal*

Similar to the provisions for groove welds outlined above, fillet weld performance qualification is valid only for materials having the same M-numbers as utilized in the

**Table 8.30** AWS B2.1 qualified positions for fillet welds

| Position(s) tested | | Position(s) qualified |
|---|---|---|
| | | Fillet in plate and pipe |
| Plate | 1 | 1 |
| | 2 | 1, 2 |
| | 3 | 1, 2, 3 |
| | 4 | 1, 2, 4 |
| | 3 + 4 | 1, 2, 3, 4 |
| Pipe | 1 | 1 |
| | 2 | 1, 2 |
| | 2 (rotated) | 1, 2 |
| | 4 | 1, 2, 4 |
| | 5 | 1, 2, 4, 5 |

*Notes*

(1) AWS appends position with the letter "F" to denote fillet weld

(2) Position of welding:

1 = Flat

2 = Horizontal

3 = Vertical

4 = Overhead

5 = Horizontal pipe

6 = 45° Inclined pipe

test weldment except as otherwise permitted in Table 8.19 (ref. AWS B2.1 5.4.1 and Table 5.3).

*Filler Metal*
Similar to the provisions for groove welds outlined above, fillet weld performance qualification is valid only for filler metals having been qualified in the test weldment as permitted in Table 8.20 (ref. AWS B2.1 5.4.2 and Table 5.5).

*Base Metal Thickness*
The base metal thickness is not restricted for welder/operator fillet weld performance qualification.

*Fillet Weld Size*
The fillet weld size is not restricted for welder/operator fillet weld performance qualification.

*Pipe Diameter*
The outside pipe or tube diameter is not restricted for welder/operator fillet weld performance qualification.

### 8.5.3 Fillet Testing and Examination Requirements for ASME IX

*Visual Examination*

There is no requirement for visual examination.

*Macro Examination*
The cut end of one of the end plate sections, approximately 1 in. (25 mm) long, in ASME IX Figure QW-462.4(b) or the cut end of one of the pipe quarter sections in ASME IX Figure QW-462.4(c), as applicable, shall be smoothed and etched with a suitable etchant (see ASME IX QW-470) to give a clear definition of the weld metal and heat-affected zone. Visual examination of the cross section of the weld metal and heat-affected zone shall reveal (ref. ASME IX QW-184):

- No incomplete fusion.
- No cracks.
- No other linear indications with a length greater than 1/32 in. [0.8 mm].
- No concavity or convexity greater than 1/16 in. [1.5 mm].
- No more than 1/8-in. [3-mm] difference between the fillet weld leg lengths.

*Break Testing*

Note ASME IX denotes break testing as fracture testing. The stem of the 4 in. [100 mm] performance specimen center section in ASME IX Figure QW-462.4(b) or the stem of the quarter section in ASME IX Figure QW-462.4(c), as applicable, shall be loaded laterally in such a way that the root of the weld is in tension. The load shall be steadily increased until the specimen fractures or bends flat upon itself (ref. ASME IX QW-182).

If the specimen fractures, the fractured surface shall show no evidence of cracks or incomplete root fusion, and the sum of the lengths of inclusions and porosity visible on the fractured surface shall not exceed 3/8 in. [10 mm] in ASME IX Figure QW-462.4(b) or 10% of the quarter section in ASME IX Figure QW-462.4(c).

*Shear Testing*

There is no optional requirement for shear testing in lieu of bend-break testing and macro examination for performance qualification using a procedure qualification.

### 8.5.4 Fillet Qualification Limits for ASME IX

In addition to the restrictions of specific variables for the different welding processes as outlined in Table 8.2 through Table 8.15, additional limitations for position, base metal, filler metal, base metal thickness, pipe diameter, and weld deposit thickness are discussed below for fillet welder/operator performance qualification for ASME IX.

*Position*

The standard welding positions for fillet welds in plate or pipe are defined in ASME IX Figure QW-461.5 and Figure QW-461.6. ASME IX defines a change in position as an angular deviation of$>\pm15$ degrees from the specified horizontal and vertical planes, and an angular deviation of$>\pm5$ degrees from the specified inclined plane are permitted during welding. This is depicted for fillet welds in ASME IX Figure QW-461.2. Test coupons welded in the specific test positions described in ASME IX Figure QW-461.5 and Figure QW-461.6 qualify the welder/operator to weld plate or pipe as permitted in Table 8.31 (ref. ASME IX Table QW-461.9).

The welder or welding operator who prepares the fillet weld procedure qualification test coupon meeting the requirements outlined in Sect. 5.2.2 is qualified to weld non-pressure retaining fillet welds only within the limits of the performance qualification. The welder or welding operator is qualified only within the limits for positions specified in Table 8.31. Qualification on a fillet weld qualifies only for fillet welding and tack welding (ref. ASME IX QW-303.2).

An organization who does production welding in a special orientation may make the tests for performance qualification in this specific orientation. Such qualifications are valid only for the flat position and for the special positions actually tested, except that an

**Table 8.31** ASME IX qualified positions for fillet welds

| Position(s) tested | | Position(s) qualified |
|---|---|---|
| | | Fillet in plate and pipe |
| Plate | 1 | 1 |
| | 2 | 1, 2 |
| | 3 | 1, 2, 3 |
| | 4 | 1, 2, 4 |
| | 3 + 4 | 1, 2, 3, 4 |
| | SP | 1, SP |
| Pipe | 1 | 1 |
| | 2 | 1, 2 |
| | 2 (rotated) | 1, 2 |
| | 4 | 1, 2, 4 |
| | 5 | 1, 2, 3, 4, 5 |
| | SP | 1, SP |

*Notes*
(1) ASME appends position with the letter "F" to denote fillet weld
(2) Position of welding:
1 = Flat
2 = Horizontal
3 = Vertical
4 = Overhead
5 = Horizontal pipe
6 = 45° Inclined pipe
SP = Special position

angular deviation of ±15 deg is permitted in the inclination of the weld axis and the rotation of the weld face, as defined in ASME IX Figures QW-461.1 and QW-461.2.

*Base Metal*
Similar to the provisions for groove welds outlined above, fillet weld performance qualification is valid only for materials having the same P-numbers as utilized in the test weldment except as otherwise permitted in Table 8.25 (ref. ASME IX QW-403.18 and QW-423).

*Filler Metal*
Similar to the provisions for groove welds outlined above, fillet weld performance qualification is valid only for materials having the same F-Numbers as utilized in the test weldment except as otherwise permitted in Table 8.26 (ref. ASME IX QW-403.18 and QW-433).

*Base Metal Thickness*

When the welder/operator fillet weld performance qualification test is performed on a pipe-to-pipe or pipe-to-plate (ref ASME IX Figure QW-462.4(c)) and the pipe outside diameter is less than 2–7/8 in. [72 mm], they are qualified for unlimited base metal thickness (ref. ASME IX QW-452.4). When the performance qualification test is conducted on the pipe-to-pipe or pipe-to-plate with an outside diameter greater than or equal to 2–7/8 in. [72 mm], or on a T-joint in plate (see. ASME IX Figure QW-462.4(b)), the qualified thickness range is shown in Table 8.32.

*Fillet Weld Size*

For welder/operator fillet weld performance qualification, the qualified fillet weld size is shown in Table 8.33.

*Pipe Diameter*

For welder/operator fillet weld performance qualification test performed on a pipe-to-pipe or pipe-to-plate (ref ASME IX Figure QW-462.4(c)) and the minimum qualified pipe outside diameter is shown in Table 8.34.

**Table 8.32** ASME IX qualified fillet weld base metal thickness

| Test weldment thickness (T) in. [mm] | Base metal thickness qualified | |
|---|---|---|
| | Minimum in. [mm] | Maximum in. [mm] |
| <3/16 [5] | T | 2T |
| ≥3/16 [5] | Unlimited | |

**Table 8.33** ASME IX qualified fillet weld size

| Test weldment thickness (T) in. [mm] | Qualified fillet weld size in. [mm] |
|---|---|
| <3/16 [5] | T maximum |
| ≥3/16 [5] | Unlimited |

**Table 8.34** ASME IX qualified fillet weld pipe diameters

| Outside diameter of test coupon in. [mm] | Diameter qualified |
|---|---|
| | Minimum in. [mm] |
| <1 [25] | Size welded |
| 1 [25] to 2–7/8 [73] | 1 [25] |
| ≥2–7/8 [73] | 2–7/8 [73] |

## 8.6 Corrosion Resistant Overlay Welder/Operator Qualification

The testing methods and the number of tests required for a corrosion resistant overlay welder/operator performance qualification test for both AWS B2.1 (Table 5.1) and ASME IX (Table QW-453) are summarized in Table 8.35.

For AWS B2.1, the bend test specimen blanks shall be removed from the locations shown in AWS B2.1 Fig. 5.12 for a plate qualification test weldment. Note that AWS B2.1 does not have provisions for corrosion resistant overlay welder/operator performance qualification on pipe. For ASME IX, bend test specimen blanks shall be removed from locations shown in ASME IX Figure QW-462.5(e) for a plate qualification test weldment and Figure QW-462.5(b) for a pipe qualification test weldment. Alternatively, ASME IX allows a welder/operator to qualify for corrosion resistant overlay with a groove weld performance qualification test when chemical composition requirements are not specified (ref. ASME IX QW-381.4).

### 8.6.1 CRO Testing and Examination Requirements for AWS B2.1

*Visual Examination*

The clad weldment shall be visually examined in accordance with AWS B2.1 5.5, Examination Procedures and Acceptance Criteria (ref. AWS B2.1 5.4.6). The appearance of the weld shall satisfy the qualifier that the welder is skilled in applying the WPS used for the test weldment (ref. AWS B2.1 5.5.1.1(2)).

*Bend Testing*

After acceptable visual examination, the clad surface shall be machined to the minimum weld metal thickness specified in the WPS. Two bend specimens are required except that 6G cladding pipe performance qualification requires three bend specimens, and the 2G cladding pipe performance qualification requires only one bend specimen (ref. AWS B2.1 5.4.6). No open discontinuity exceeding 1/16 in. [1.5 mm] measured in any direction on the surface shall be permitted in the cladding, and no open defects exceeding 1/8 in. [3 mm] in length shall be permitted at the weld interface after bending (ref. AWS B2.1 5.5.3.3).

**Table 8.35** Test methods required for corrosion resistant overlay performance qualification

| | AWS B2.1 | ASME IX |
|---|---|---|
| Visual examination | Yes | Not required |
| Bend testing | Yes (qty.1,2,or3) | Yes (qty.2) |

### 8.6.2 CRO Qualification Limits for AWS B2.1

In addition to the restrictions of specific variables for the different welding processes as outlined in Table 8.2 through Table 8.15, additional limitations for Position, Base Metal, Filler Metal, Base Metal Thickness, Pipe Diameter, and Weld Deposit Thickness are discussed below for corrosion resistant overlay welder/operator performance qualification for AWS B2.1.

*Position*
The standard welding positions for corrosion resistant overlay on plate or pipe are defined in AWS B2.1 Figure A.1A. AWS B2.1 defines a change in position as an angular deviation of>±15 degrees from the specified horizontal and vertical planes, and an angular deviation of>±5 degrees from the specified inclined plane are permitted during welding. This is depicted for corrosion resistant overlay in AWS B2.1 Figure A.1B.

Test coupons welded in the specific test positions described in AWS B2.1 Annex A1.1 qualify the welder to weld plate or pipe as permitted in Table 8.36 (ref. AWS B2.1 Table 5.5). Cladding qualification qualifies only for cladding (ref. AWS B2.1 5.1.15). For surfacing applications, qualification on plate qualifies for plate only except that qualification on plate in the flat position also qualifies on pipe in the flat position. Qualification on pipe in any position shown above for cladding or hardfacing also qualifies for plate in the positions allowed in the table (ref. AWS B2.1 Table 5.5 Note e). Weldment orientation other than the specific test positions shown in Annex A is permitted, but such tests qualify only for the orientation tested (ref. AWS B2.1 5.4.3).

*Base Metal*
Similar to the provisions for groove welds outlined above, corrosion resistant overlay performance qualification is valid only for materials having the same M-numbers as utilized in the test weldment except as otherwise permitted in Table 8.18 (ref. AWS B2.1 5.4.1 and Table 5.3).

*Filler Metal*
Similar to the provisions for groove welds outlined above, corrosion resistant overlay performance qualification is valid only for filler metals having been qualified in the test weldment as permitted in Table 8.20 (ref. AWS B2.1 5.4.2 and Table 5.5).

*Base Metal Thickness*
The base metal thickness is not restricted for welder/operator corrosion resistant overlay performance qualification.

*Weld Deposit Thickness*
The weld deposit thickness is not restricted for welder/operator corrosion resistant overlay performance qualification.

**Table 8.36** AWS B2.1 qualified positions for corrosion resistant overlay

| Position(s) tested | | Position(s) qualified |
|---|---|---|
| | | Fillet in plate and pipe |
| Plate or plate | 1 | 1 |
| | 2 | 1, 2 |
| | 3 | 1, 3 |
| | 4 | 1, 4 |
| | 3 + 4 | 1, 3, 4 |
| | 2 + 3 + 4 | 1, 2, 3, 4, 5, 6 |
| | 5 | 1, 3, 4 |
| | 6 | 1, 2, 3, 4, 5, 6 |

*Notes*
(1) AWS appends position with the letter "C" to denote surfacing welds
(2) Position of welding:
1 = Flat
2 = Horizontal
3 = Vertical
4 = Overhead
5 = Horizontal pipe
6 = 45° Inclined pipe

*Pipe Diameter*
The outside pipe or tube diameter is not restricted for welder/operator corrosion resistant overlay performance qualification.

### 8.6.3 CRO Testing and Examination Requirements for ASME IX

*Visual Examination*

There is no requirement for visual examination.

*Bend Testing*
ASME IX requires performance qualification testing by 2 transverse side bends (ref. ASME IX Table QW-453). The preparation and dimensions of specimen blanks for bend testing are provided in ASME IX QW-161.1 and Figure QW-462.2 for transverse side bend specimens. The test specimens shall be cut so that there are two specimens perpendicular to the direction of the welding. For coupons that are less than 3/8 in. [10 mm] thick, the width of the side-bend specimens may be reduced to the thickness of the test coupon (ref. ASME IX QW-214.2(b)). Bend specimens shall be bent in one of the

guided bend test fixtures shown in ASME IX Figure QW-466.1 test jig dimensions, Figure QW-466.2 guided-bend roller jig, or Figure QW-466.3 guided-bend wrap around jig.

The acceptance criteria for side bend testing in ASME IX QW-163 states:

- There shall be no open discontinuity exceeding 1/16 in. [1.5 mm] in the cladding measured in any direction.
- There shall be no open discontinuity exceeding 1/8 in. [3 mm] along the approximate weld interface.

### 8.6.4 CRO Qualification Limits for ASME IX

In addition to the restrictions of specific variables for the different welding processes as outlined in Table 8.2 through Table 8.15, additional limitations for Position, Base Metal, Filler Metal, Base Metal Thickness, Pipe Diameter, and Weld Deposit Thickness are discussed below for corrosion resistant overlay welder/operator performance qualification for ASME IX.

*Position*

Since position is defined as an essential variable for the common arc welding process (see Sect. 8.3.5, Tables 8.9 and 8.10), the limitations of positions qualified for groove welds shall apply to corrosion resistant overlay welds (ref. ASME IX QW-381.2(c)). These limitations and the qualified positions are the same as those for groove welds defined in Sect. 8.4.4 and Table 8.24. Welders or welding operators who pass the tests for corrosion-resistant weld metal overlay cladding shall only be qualified to apply corrosion-resistant weld metal overlay portion of a groove weld joining clad materials or lined materials (ref. ASME IX QW-381.2 (b).

*Base Metal*

Similar to the provisions for groove welds outlined above, corrosion resistant overlay performance qualification is valid only for materials having the same P-numbers as utilized in the test weldment except as otherwise permitted in Table 8.25 (ref. ASME IX QW-403.18 and QW-423).

*Filler Metal*

Similar to the provisions for groove welds outlined above, corrosion resistant overlay performance qualification is valid only for materials having the same F-Numbers as utilized in the test weldment except as otherwise permitted in Table 8.26 (ref. ASME IX QW-403.18 and QW-433).

*Base Metal Thickness*
The base metal thickness is not restricted for welder/operator corrosion resistant overlay performance qualification.

*Weld Deposit Thickness*
The weld deposit thickness is not restricted for welder/operator corrosion resistant overlay performance qualification (ref. ASME IX QW-381.2(c)).

*Pipe Diameter*
Since pipe diameter is defined as an essential variable for the common arc welding process (see Sect. 8.3.3 and Table 8.6), the limitations of pipe diameter qualified for groove welds also apply to corrosion resistant overlay welds (ref. ASME IX QW-381.2(c)). These limitations are the same as those for groove welds defined in Table 8.28. The limitations on diameter qualified shall apply only to welds deposited in the circumferential direction (QW-381.2(c)).

## 8.7 Hardfacing Overlay Welder/Operator Qualification

The testing methods and the number of tests required for a hardfacing overlay welder/operator performance qualification test for both AWS B2.1 (Table 5.1) and ASME IX (Table QW-453) are summarized in Table 8.37.

For AWS B2.1, the macro examination specimen blanks shall be removed from the locations shown in AWS B2.1 Fig. 5.13 for a plate qualification test weldment. Note that AWS B2.1 does not have provisions for hardfacing welder/operator performance qualification on pipe. For ASME IX, the macro examination specimen blanks shall be removed from locations shown in ASME IX Figure QW-462.5(d) for a plate qualification test weldment and Figure QW-462.5(c) for a pipe qualification test weldment. There are no provisions in AWS B2.1 or ASME IX for welder/operator hardfacing performance qualification using special test weldments.

### 8.7.1 HFO Testing and Examination Requirements for AWS B2.1

*Visual Examination*

**Table 8.37** Test methods required for hardfacing overlay performance qualification

| | AWS B2.1 | ASME IX |
|---|---|---|
| Visual examination | Yes | Not required |
| Liquid penetrant | Not required | Yes |
| Macro examination | Yes | Yes |

Prior to removing the macro examination specimen blanks identified, the hardfaced surface shall be visually examined in accordance with the referencing document (ref. AWS B2.1 5.4.7).

*Liquid Penetrant Examination for AWS B2.1*
There is no requirement for liquid penetrant examination.

*Macro Examination*
Unless otherwise specified in the referencing document one transverse macro shall be removed as shown in AWS B2.1 Fig. 5.13, and the weld shall show complete fusion (ref. AWS B2.1 5.4.7).

### 8.7.2 HFO Qualification Limits for AWS B2.1

In addition to the restrictions of specific variables for the different welding processes as outlined in Table 8.2 through Table 8.15, additional limitations for Position, Base Metal, Filler Metal, Base Metal Thickness, Pipe Diameter, and Weld Deposit Thickness are discussed below for hardfacing overlay welder/operator performance qualification for AWS B2.1.

*Position*
The qualified positions for hardfacing overlay performance qualification are the same as those for corrosion resistant overlay performance qualification outlined in Table 8.36.

*Base Metal*
Similar to the provisions for groove welds outlined above, hardfacing overlay performance qualification is valid only for materials having the same M-numbers as utilized in the test weldment except as otherwise permitted in Table 8.18 (ref. AWS B2.1 5.4.1 and Table 5.3).

*Filler Metal*
Similar to the provisions for groove welds outlined above, hardfacing overlay performance qualification is valid only for filler metals having been qualified in the test weldment as permitted in Table 8.20 (ref. AWS B2.1 5.4.2 and Table 5.5).

*Base Metal Thickness*
The base metal thickness is not restricted for welder/operator hardfacing overlay performance qualification.

*Fillet Weld Size*
The fillet weld size is not restricted for welder/operator hardfacing overlay performance qualification.

*Pipe Diameter*
The outside pipe or tube diameter is not restricted for welder/operator hardfacing overlay performance qualification.

### 8.7.3 HFO Testing and Examination Requirements for ASME IX

*Visual Examination*

There is no requirement for visual examination.

*Liquid Penetrant Examination for ASME IX*
At a thickness greater than or equal to the minimum thickness specified in the WPS, the weld surface shall be examined by the liquid penetrant method. Surface conditioning prior to liquid penetrant examination is permitted (ref. ASME IX QW-382.1(c)). The examination shall meet the following acceptance requirements, or as specified in the WPS (ref ASME IX QW-195.2):

- No linear indications with major dimensions greater than 1/16 in. [1.5 mm] having a length greater than three times the width.
- No rounded indications of circular or elliptical shape greater than 3/16 in. [5 mm] with the length equal to or less than three times the width.
- No more than four rounded indications of circular or elliptical shape with the length equal to or less than three times the width in a line separated by 1/16 in. [1.5 mm] or less (edge-to-edge).

*Macro Examination*
The base metal shall be sectioned transversely to the direction of the hard-facing overlay. The two faces of the hard facing exposed by sectioning shall be polished and etched with a suitable etchant and shall be visually examined with 5X magnification for cracks in the base metal or the heat-affected zone, lack of fusion, or other linear defects. The overlay and base metal shall meet the requirements specified in the WPS. All exposed faces shall be examined (ref. ASME IX QW-382.1(b)).

### 8.7.4 HFO Qualification Limits for ASME IX

In addition to the restrictions of specific variables for the different welding processes as outlined in Table 8.2 through Table 8.15, additional limitations for Position, Base Metal Thickness, and Weld Deposit Thickness are discussed below for hardfacing overlay welder/operator performance qualification for ASME IX.

*Position*
Since position is defined as an essential variable for the common arc welding process (see Sect. 8.3.5, Tables 8.9 and 8.10), the limitations of positions qualified for groove welds shall apply to hardfacing overlay welds (ref. ASME IX QW-382.1(f)). These limitations and the qualified positions are the same as those for groove welds defined in Sect. 8.4.4 and Table 8.24.

*Base Metal*
Similar to the provisions for groove welds outlined above, hardfacing overlay performance qualification is valid only for materials having the same P-numbers as utilized in the test weldment except as otherwise permitted in Table 8.25 (ref. ASME IX QW-403.18 and QW-423).

*Filler Metal*
Similar to the provisions for groove welds outlined above, hardfacing overlay performance qualification is valid only for materials having the same F-Numbers as utilized in the test weldment except as otherwise permitted in Table 8.26 (ref. ASME IX QW-403.18 and QW-433). In addition, qualification with one AWS classification within an SFA specification qualifies for all other AWS classifications (ref. ASME IX QW-382.1(g)) in that SFA specification.

*Base Metal Thickness*
The base metal thickness is not restricted for welder/operator hardfacing overlay performance qualification.

*Weld Deposit Thickness*
The weld deposit thickness is not restricted for welder/operator corrosion resistant overlay performance qualification (ref. ASME IX QW-382.1(f)).

#### 8.7.4.1 Pipe Diameter

Since pipe diameter is defined as an essential variable for the common arc welding process (see Sect. 8.3.3 and Table 8.6), the limitations of pipe diameter qualified for groove welds also apply to hardfacing overlay welds (ref. ASME IX QW-382.1(f)). These limitations are the same as those for groove welds defined in Table 8.28. The limitations on diameter qualified shall apply only to welds deposited in the circumferential direction (QW-382.1(f)).

## 8.8 Retests, Revocation, Continuity, Expiration and Renewal of Qualification

The following sections discuss the provisions for dealing with a failed, revoked, or expired welder/operator performance qualification test.

### 8.8.1 Retesting

A welder or welding operator who fails any requirement of a performance qualification test may be retested under the following conditions. Both AWS B2.1 and ASME IX allow the welder/operator to be immediately retested without further training or practice. This so called "two-for-one" provision requires the welder/operator to complete two qualification test weldments for each position failed (ref. AWS B2.1 5.1.10(1) and ASME IX QW-321). If after a failed qualification the welder/operator receives further training and/or practice, only one qualification test weldment is required for each position failed. The requirements for satisfactory training and/or practice is somewhat ambiguous but subject to judgement by the weld test supervisor for both AWS B2.1 and ASME IX. ASME IX has further provisions depending on the testing and examination requirements specifically for visual examination, mechanical testing, or volumetric NDE.

### 8.8.2 Revocation of Qualification

Both AWS B2.1 and ASME IX have provisions to revoke a welder/operator's performance qualification(s) when there is a specific reason to question their ability to make welds that meet specifications (ref. AWS B2.1 5.1.11(3) and ASME IX QW-322.2). This only affects the specific performance qualifications supporting the welding be performed and does not affect all other performance qualifications. Once revoked, requalification is permitted utilizing a test coupon appropriate for the revoked qualification(s) along with all applicable qualification variables, examination, and testing requirements (ref. AWS B2.1 5.1.11(4) and ASME IX QW-322.3). Successful completion of the qualification test restores the revoked qualification(s). Requalification may be performed on a workmanship or production test coupon.

### 8.8.3 Continuity and Expiration of Qualification

Both AWS B2.1 and ASME IX have provisions for continuity that require a welder or welding operator to utilize a welding process for which they are certified within a six month period of certification or recertification, otherwise the performance qualification for that process expires. This continuity of performance qualification shall be confirmed and tracked by the qualifying organization. A welder/operator may renew qualification for the expired process by performing a single test weldment that meets the requirements of the previous qualification which will reinstate their previous qualification for that process, material thicknesses, diameters, positions, and all other qualification variables (ref. AWS B2.1 5.1.1 and ASME IX QW-322.1). Requalification may be performed on a workmanship or production test coupon.

## 8.9 Summary

This chapter discussed the requirements for the performance qualification of welders and welding operators in accordance with the rules of AWS B2.1 and ASME IX. The required welding variables to be documented in the performance qualification test record are discussed for each arc welding process. The required examination and test methods along with their acceptance criteria are also outlined. Finally, the qualified limits and ranges are discussed.

The manufacturer's authorised representative in the EU is Springer Nature Customer Service Centre GmbH, Europaplatz 3, 69115 Heidelberg, Germany. If you have any concerns regarding our products, please contact ProductSafety@springernature.com

Printed and bound by CPI Group (UK) Ltd, Croydon, CR0 4YY

30/09/2026

02225561-0020